SCIENCE
A CLOSER LOOK

BUILDING SKILLS

Reading and Writing Workbook

caItlin Lueker

A

The McGraw-Hill Companies

Macmillan
McGraw-Hill

Published by Macmillan/McGraw-Hill, of McGraw-Hill Education, a division of The McGraw-Hill Companies, Inc.,
Two Penn Plaza, New York, New York 10121.

...yright © by Macmillan/McGraw-Hill. All rights reserved. No part of this publication may be reproduced or
...uted in any form or by any means, or stored in a database or retrieval system, without the prior written consent
...Graw-Hill Companies, Inc., including, but not limited to, network storage or transmission, or broadcast for
...earning.

...e United States of America

...1 20

Contents

LIFE SCIENCE

© Macmillan/McGraw-Hill

Contents

© Macmillan/McGraw-Hill

Contents

EARTH SCIENCE

© Macmillan/McGraw-Hill

Contents

© Macmillan/McGraw-Hill

Contents

PHYSICAL SCIENCE

© Macmillan/McGraw-Hill

Contents

© Macmillan/McGraw-Hill

Name ______________________ Date __________

LESSON
Outline

What Living Things Need

Use your book to help you fill in the blanks.

What do living things need?

1. All living things have needs they must meet in order to grow and ____________ .
2. Most animals need to move, ____________ air, drink water, and eat food to grow.
3. Plants are ____________ things, too.
4. Plants also change and ____________ over time.
5. Plants need ____________ , water, and space to grow.
6. Plants use their parts to make their own ____________ .

© Macmillan/McGraw-Hill

Name ______________________ Date __________

How do plants make food?

7. Plants use ____________, air, water, and minerals to make their own food.

8. Minerals come from tiny bits of ____________ and rocks.

9. When plants make their own food, they also make a ____________ called oxygen.

10. People and animals need ____________ to breathe.

Critical Thinking

11. How do the parts of a plant help it get what it needs to live?

____

____

____

____

____

© Macmillan/McGraw-Hill

Name ______________________ Date ____________

What Living Things Need

Choose a word from the box to answer each riddle.

leaves	oxygen	stem
minerals	roots	

1. I can be found in the ground. I am made from tiny bits of rocks and soil. What am I?

____

2. I help a plant take in air and sunlight. What am I?

____

3. Food and water travel through me to get to all parts of the plant. What am I?

____

4. I help a plant take in minerals from the ground. What am I?

____

5. You breathe me in so you can live. What am I?

____

© Macmillan/McGraw-Hill

LESSON
Cloze Activity

Name ______________________ Date __________

What Living Things Need

Fill in the blanks. Use the words from the box.

gas	minerals	roots	sunlight
leaves	oxygen	stem	

Plants, animals, and people all need food, air, and water to live. Plants need ______________ and space to grow, too.

Animals and people must move around to get what they need, but plants have parts that help them survive in one place. The ______________ hold the plant in the ground. They also take in ______________ from soil.

The ______________ take in air and sunlight to make food. Food and water travel through the ______________ to reach all parts of the plant. When plants make food, they give off ______________ into the air.

Oxygen is a ______________ that we need to survive. We can find oxygen in the air we breathe and the water we drink.

© Macmillan/McGraw-Hill

Name ______________________ Date ____________

LESSON
Outline

Plants Make New Plants

Use your book to help you fill in the blanks.

Where do seeds come from?

1. A ______________ is a special plant part that can grow into a new plant.

2. Seeds are made inside a ______________ .

3. Sometimes a flower will ______________ seeds inside of a fruit.

4. Flowers also make ______________ , the sticky powder that helps them make seeds.

5. Bugs and ______________ can help move pollen from flower to flower.

6. Wind and ______________ from rain can move pollen, too.

How do seeds look?

7. Seeds can have many ______________ and shapes, just like plants.

© Macmillan/McGraw-Hill

Name ______________________ Date __________

8. All seeds have seed ____________ or fruit to protect them as they grow.

How do seeds grow?

9. The ____________ of a plant begins with a seed.

10. The way plants grow, live, and ____________ is called their life cycle.

11. Most seeds need ____________ , water, food, and a little heat to become new plants.

12. A new plant has the same life cycle as its

    ____________ plant.

Critical Thinking

13. How are new plants that grow from seeds like their parent plants?

____

____

____

____

© Macmillan/McGraw-Hill

Name ______________________________ Date ____________

LESSON
Vocabulary

Plants Make New Plants

Read the sentences below. Write TRUE if the sentence is true. Write NOT TRUE if the sentence is false.

1. ________________ Inside a seed, there is a sticky powder called pollen.

2. ________________ Part of a flower can turn into fruit.

3. ________________ The fruit protects the seeds inside it.

4. ________________ A life cycle shows how a plant grows, lives, and dies.

5. ________________ An adult plant can grow into a seedling.

6. ________________ Seeds have a special coat that keeps them from drying out.

© Macmillan/McGraw-Hill

LESSON
Cloze Activity

Name ______________________ Date ____________

Plants Make New Plants

Fill in the blanks. Use the words from the box.

flowers	life cycle	seed coat	seeds
fruit	pollen	seedling	

Plants make new plants during their life cycle. A ______________ shows how a living thing grows, lives, and dies. The life cycle of a plant begins with a seed. A special covering called a ______________ helps protect the seed. The seed sprouts a ______________ if it gets enough food, water, and heat. It may grow ______________ as it becomes an adult plant.

A sticky material called ______________ is found inside of flowers. Flowers use pollen to make seeds. Part of the flower can also grow into a fruit that has ______________. When the ______________ becomes ripe, it falls to the ground. Then the seeds can turn into new plants.

© Macmillan/McGraw-Hill

Name ______________________________ Date ____________

Main Idea and Details

Write About It

On a separate piece of paper, write a paragraph about a flower that you observed. Include a main idea and details.

Getting Ideas

Write the name of a flower in the Main Idea oval. Write a detail about the flower in each detail oval.

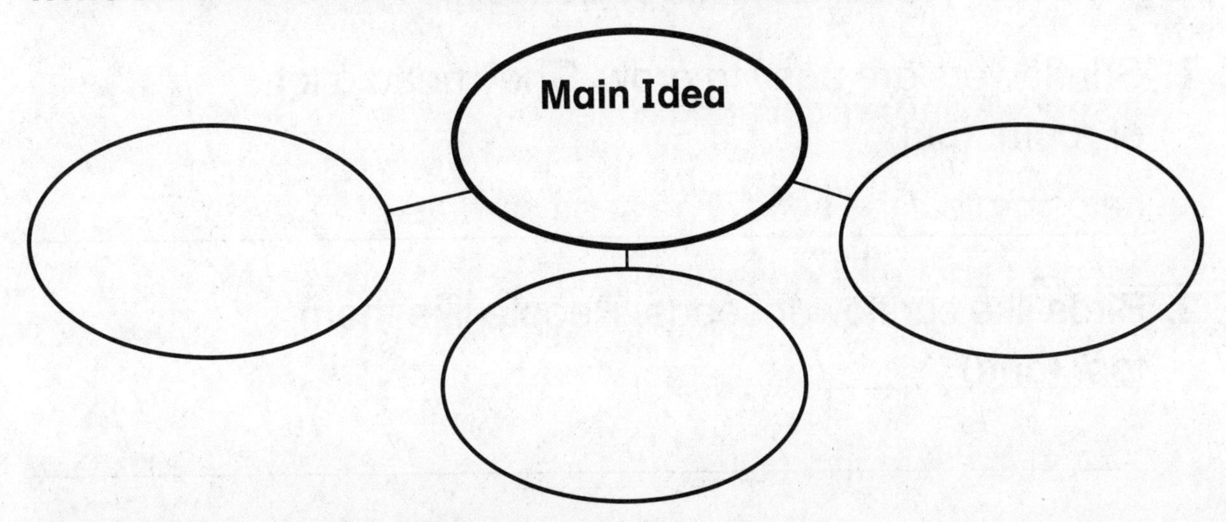

Planning and Organizing

Hector wrote three sentences about a sunflower. Write Detail if the sentence tells a detail. Write Main Idea if the sentence tells the main idea.

1. ______________ Birds like sunflower seeds.
2. ______________ A sunflower has seeds.
3. ______________ A sunflower is useful.

© Macmillan/McGraw-Hill

Name ______________________ Date __________

Drafting

Write a sentence that tells the main idea about your flower.

____

On a separate piece of paper, write a whole paragraph. Give details about your flower.

Revising and Proofreading

Hector wrote some sentences. Use the words in parentheses ( ) to combine his sentences.

1. Sunflowers are easy to grow. They need a lot of room. (but)

   ____

2. Birds like sunflower seeds. People like them, too. (and)

   ____

3. The seeds are very healthful. They make a good snack. (so)

   ____

Now revise and proofread your writing. Ask yourself:

- ▶ Did I include the main ideas and details?
- ▶ Did I correct all mistakes?

© Macmillan/McGraw-Hill

Name ______________________________ Date ____________

LESSON **Outline**

How Plants Are Alike and Different

Use your book to help you fill in the blanks.

How are plants like their parents?

1. Oak trees make ______________ that grow into new oak trees.
2. Sunflowers make seeds that grow to look just like their ______________ sunflower.
3. A ______________ is a way a living thing looks or acts like its parent.
4. Some plants and animals share many ______________ with their parents.
5. Other plants and animals ______________ just a few traits with their parents.

How can plants survive in different places?

6. Plants can ______________ to get what they need from their environment.

© Macmillan/McGraw-Hill

Name ______________________ Date ____________

7. Some plants in forests grow large ____________ that help them get more sunlight.

8. Plants that live in dry places grow thick

____________ to store water.

9. Plants can change to stay ____________ in their environment, too.

10. Some plants grow in ways that keep away

____________ that want to eat them.

11. Other plants change to stay safe from

____________ where they live.

Critical Thinking

12. What do you think would happen to a plant that did not change to fit in its environment? Why?

____

____

____

____

____

© Macmillan/McGraw-Hill

Name ______________________________ Date ____________

How Plants Are Alike and Different

Write the correct word for each sentence. Then find and circle the word in the puzzle below.

1. The ways plants and animals look and act like their parents are called ______________ .
2. Plants can ______________ to fit the place they live.
3. When a seed sprouts, the ______________ always grow down.
4. Plants do not pass some traits down to their ______________ .

E	X	R	Q	F	J	L	M	S	N	T
U	S	C	H	A	N	G	E	Q	B	N
L	F	M	W	O	S	D	V	L	S	U
M	Y	E	S	L	T	R	A	I	T	S
R	O	O	T	S	H	O	W	E	K	N
A	X	A	B	K	J	L	Z	N	E	Y
B	Z	O	F	F	S	P	R	I	N	G

© Macmillan/McGraw-Hill

LESSON
Cloze Activity

Name ____________________ Date __________

How Plants Are Alike and Different

Fill in the blanks. Use the words from the box.

dry	offspring	touches	safe
change	parents	trait	

Some people in your family probably look alike. They may even act alike! Plants can look and act like their ______________, too. A ______________ is a way a plant or animal looks or acts like its parent. One kind of trait that a plant could share with its ______________ is the shape of its leaves.

All plants are also alike in that they can ______________ to fit the place they live. Sometimes plants change in order to stay ______________ from the weather. Plants that live in ______________ places can store water in their thick stems. Plants may also change to stay safe from animals. Some plants can even change when an animal ______________ them! Venus flytraps are plants that trap and eat bugs that wander onto their leaves.

© Macmillan/McGraw-Hill

Name ______________________ Date ____________

The Power of Periwinkle

Read the Reading in Science pages in your book. Use what you read to make inferences based on the sentences in the "What I Know" column. Write your inferences on the chart.

What I Know	What I Infer
People who live in forests all over the world know about helpful plants.	
The rosy periwinkle was first found in the forests of Madagascar.	
Scientists study plants in forests all over the world.	

© Macmillan/McGraw-Hill

Name ______________________ Date ____________

1. What did you learn about how people in Madagascar use rosy periwinkle?

____

____

2. Look at the picture of rosy periwinkle in your book. Draw a picture of it. Then write your own caption.

Caption: ______________________________

____

Write About It

Predict. What might happen if scientists find more helpful plants in the forests of the world?

____

____

© Macmillan/McGraw-Hill

Name ______________________ Date ____________

Plants

Fill in the blanks. Write the words in the puzzle.

Down

1. The ways that plants and animals look like their parents are called ______________.
3. The sticky powder inside a flower is called ______________.
5. A ______________ is the part of a plant that can grow into a new plant.

Across

2. The ______________ holds up the plant.
4. When plants make food, they give off ______________.

1. 3. 4. 5. 2.

© Macmillan/McGraw-Hill

Name ______________________ Date ____________

Match the words in the box to the pictures below.

flower	leaves	roots	seedling

1. 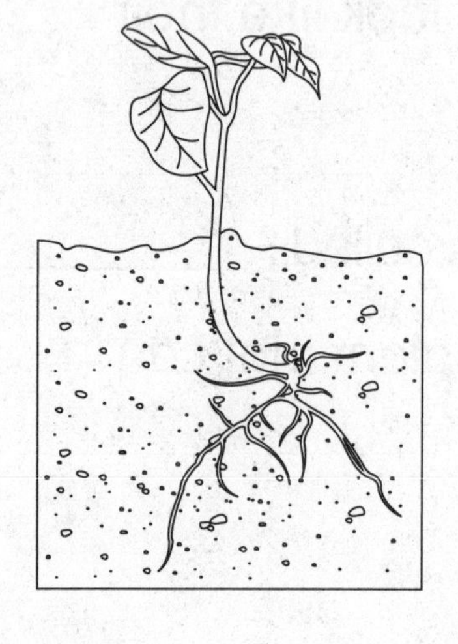

____

2. 

____

3. 

____

4. 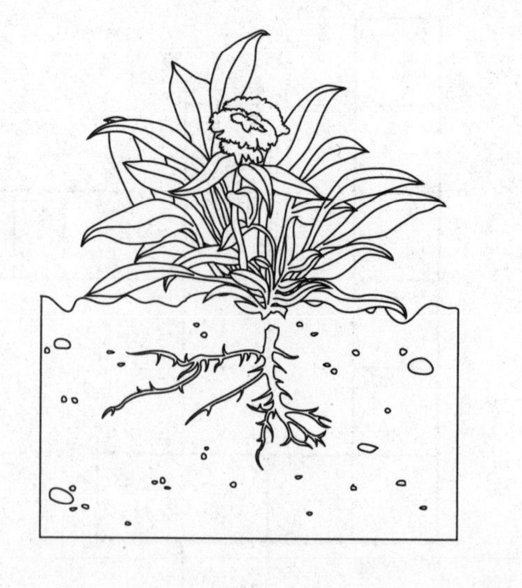

____

© Macmillan/McGraw-Hill

Name ______________________ Date ____________

CHAPTER
Concept Map

Animals

Fill in the important ideas as you read the chapter. Some ideas have already been filled in for you.

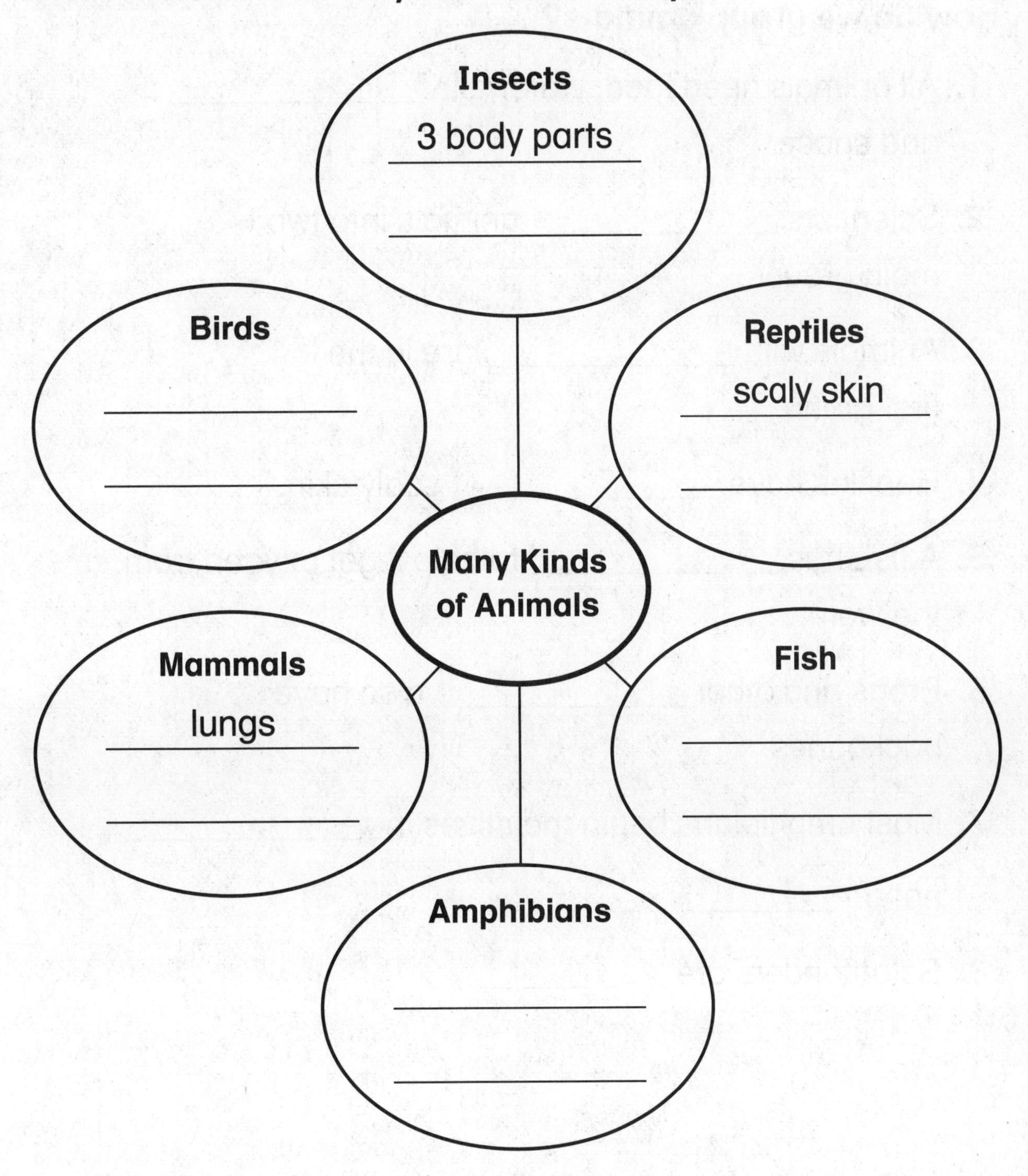

© Macmillan/McGraw-Hill

Name ______________________________ Date ____________

Animal Groups

Use your book to help you fill in the blanks.

How do we group animals?

1. All animals need food, water, air, ______________, and space.
2. Scientists ______________ animals into two main groups.
3. Animals with ______________ are in the first group.
4. Reptiles have ______________, scaly skin.
5. A fish has ______________ to help it get oxygen from the water.
6. Frogs and other ______________ also have backbones.
7. Most amphibians begin their lives in ______________, not on ______________.
8. Salamanders are ______________.

© Macmillan/McGraw-Hill

9. Birds have ______________ and lay eggs.

10. Mammals have ______________ or hair, and birds have feathers.

What are some animals without backbones?

11. Some animals without backbones grow coverings like ______________ to keep them safe.

12. Insects have ______________, six legs, and no backbone.

13. Jellyfish are soft. They ______________ other animals to stay safe.

Critical Thinking

14. How are a bird and an insect alike? How are they different?

____

____

____

____

____

____

© Macmillan/McGraw-Hill

Name ____________________ Date __________

Animal Groups

Label each animal with its animal group. Use the words in the box.

amphibian	fish	mammal
bird	insect	reptile

1. 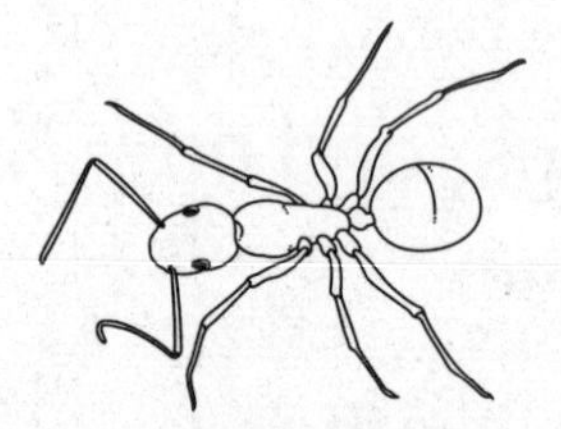

____

4. 

____

2. 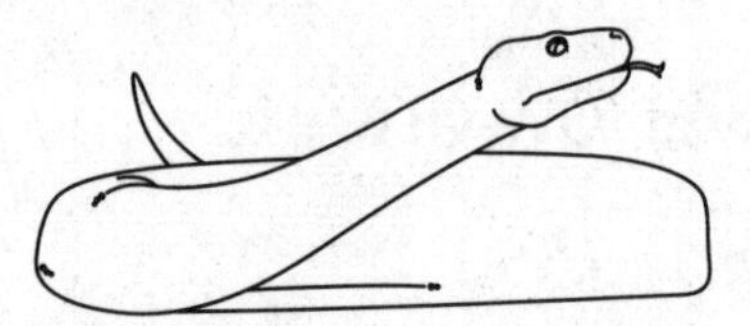

____

5. 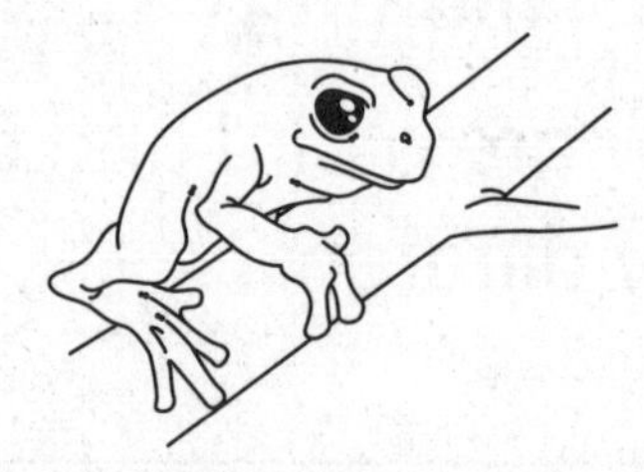

____

3. 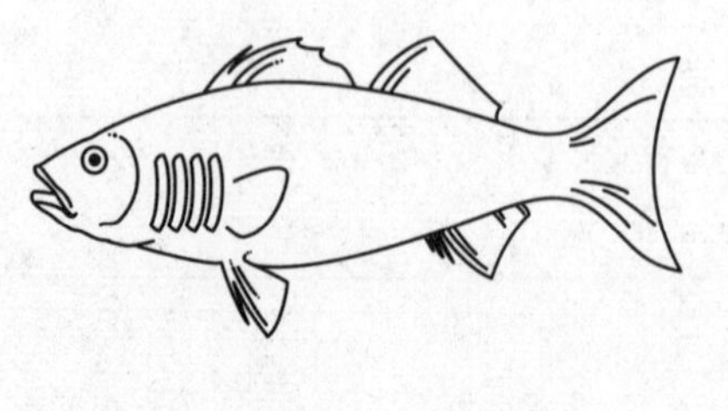

____

6. 

____

© Macmillan/McGraw-Hill

Name ______________________ Date __________

Animal Groups

Fill in the blanks. Use the words from the box.

amphibian	bird	fish	mammal
backbone	classify	insect	reptile

Our world is home to many kinds of animals. When scientists study animals, they ______________ them into two groups. The groups are animals with a ______________ and animals without a backbone. Birds, ______________ , mammals, reptiles, and amphibians all have a backbone.

A ______________ is the only animal that has feathers. All birds have two wings, but not all birds can fly. Fish use ______________ to help them get oxygen from the water where they live. An ______________ has moist skin to help it live on land and in water. A ______________ has dry, scaly skin to protect it and keep it warm. A ______________ has fur and hair to keep it warm. Reptiles and mammals use their lungs to get oxygen.

© Macmillan/McGraw-Hill

Name ______________________ Date ____________

Animals Grow and Change

Use your book to help you fill in the blanks.

What is a life cycle?

1. A ______________ tells how an animal begins life, lives, and dies.
2. Insects, birds, fish, reptiles, and ______________ start their life cycle as eggs.
3. The life cycle of a ______________ starts when it is born as a ______________ baby.

What are some other animal life cycles?

4. Some ______________ do not look like their parents at all when they are young.
5. Animals such as butterflies, frogs, and ______________ change during their lives.
6. A caterpillar is the ______________ that hatches from a butterfly egg.

© Macmillan/McGraw-Hill

7. A caterpillar enters the ______________ stage when it is time to turn into a butterfly.

8. During this stage, the caterpillar's ______________ becomes a hard shell.

9. Soon, an adult ______________ comes out of the shell and flies away.

Critical Thinking

10. How does a human change during its life cycle?

____

____

____

____

____

____

____

____

____

____

____

© Macmillan/McGraw-Hill

LESSON
Vocabulary

Name ______________________________ Date ____________

Animals Grow and Change

Write the correct word next to each stage of this butterfly's life cycle.

butterfly	larva
egg	pupa

1.

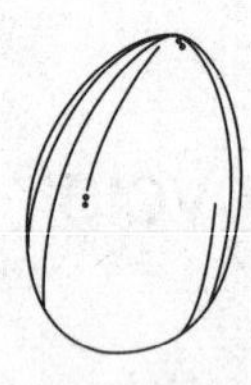

This animal begins as an ____________ .

2.

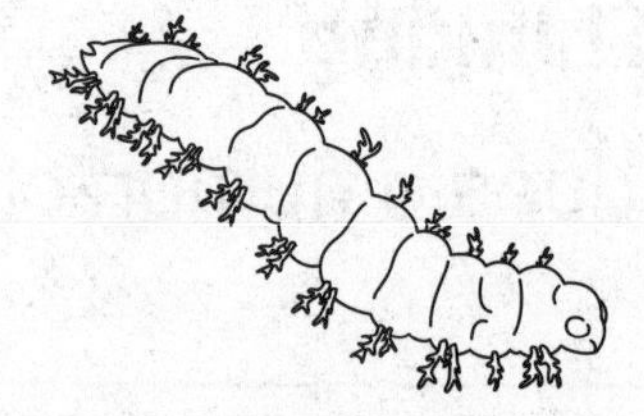

When it hatches, a ____________ comes out. This is called a caterpillar.

3.

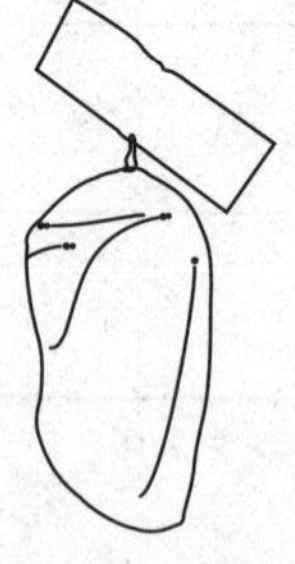

The caterpillar's skin becomes a hard shell. This is called the ____________ stage.

4.

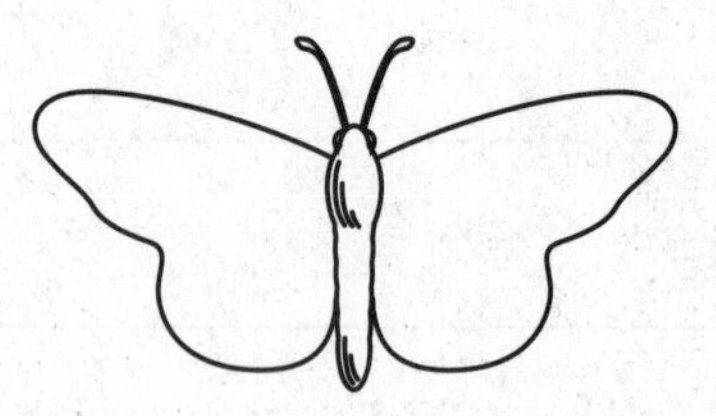

Soon, an adult ____________ comes out of the shell.

© Macmillan/McGraw-Hill

Name ______________________________ Date ____________

LESSON
Cloze Activity

Animals Grow and Change

Fill in the blanks. Use the words from the box.

butterfly	larva	mammals	pupa
egg	life cycle	older	shell

Animals begin their lives in different ways. A ______________ shows how an animal starts life, grows to be an adult, and dies.

Most ______________ begin their lives when they are born as live young. As they grow ______________ , they look more like their parents.

Many insects begin life differently. A ______________ begins life as an egg. When the ______________ hatches, a ______________ comes out. Soon, the larva stops moving and forms a hard ______________ . This is called the ______________ stage. Finally, a colorful butterfly comes out. It waits for its wings to dry and then flies away.

© Macmillan/McGraw-Hill

Name ______________________ Date ____________

Meet Nancy Simmons

Read the Reading in Science pages in your book. Look for the main idea and details as you read. Remember, the main idea is the most important idea in the passage. Write the main idea in the chart below. Be sure to also write any details that help give more information about the main idea.

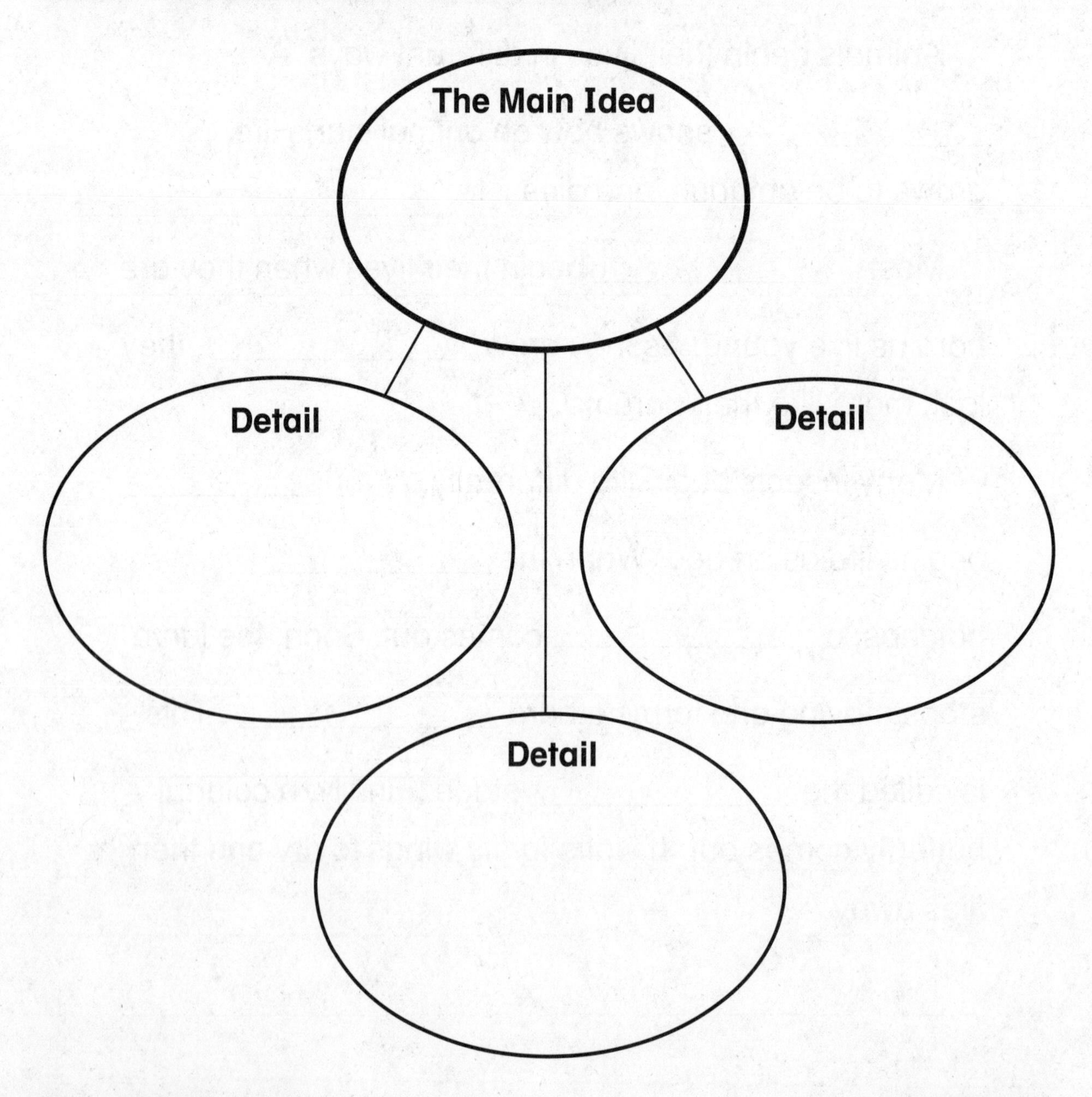

© Macmillan/McGraw-Hill

Name ____________________ Date __________

1. What did you learn about the false vampire bat? How did you learn it?

2. What are baby bats called? What did you learn about how a young bat looks just after it is born?

Write About It

Find the Main Idea. How is a pup different from an adult bat? Use the chart you made to help you write your answer.

© Macmillan/McGraw-Hill

LESSON **Outline**

Name ______________________________ Date ____________

Staying Alive

Use your book to help you fill in the blanks.

Why do animals act and look the way they do?

1. Animals can ______________ , or adapt, to help them stay alive.

2. An ______________ is a body part or a way an animal acts that helps it stay alive.

3. The long neck of a ______________ is an adaptation.

4. The adaptation helps the giraffe ______________ leaves from the tops of trees.

5. Some adaptations, like ______________ , help animals hide from other animals.

6. Camouflage can be a color or a body ______________ that helps an animal hide in nature.

7. A ptarmigan is a ______________ that has brown feathers in the summer.

8. In the winter, the ptarmigan's feathers turn ______________ so it can blend in with the snow.

© Macmillan/McGraw-Hill

Name ______________________________ Date ____________

How do animals stay safe?

9. Some animals move in large ______________ to stay safe.

10. Staying together in a large group helps ______________ smaller fish from being eaten by bigger fish.

11. Some animals, like bears and mice, ______________ during the cold winter.

12. Other animals ______________ to places where they can find food and stay warm during winter.

Critical Thinking

13. What adaptations does a bear have to help it stay safe?

____

____

____

____

____

____

____

© Macmillan/McGraw-Hill

Name ______________________ Date ____________

Staying Alive

Describe each animal's adaptations to stay alive.

1.

giraffe

____

____

____

____

2.

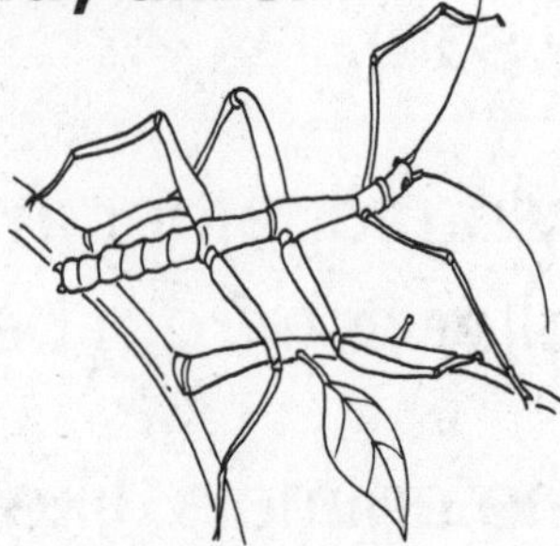

stick bug

____

____

____

____

3.

zebra

____

____

____

____

4.

hawk

____

____

____

____

© Macmillan/McGraw-Hill

Name ______________________________ Date ____________

Staying Alive

Fill in the blanks. Use the words from the box.

adaptation	camouflage	groups	shape
blend	color	pattern	winter

There are many ways in which animals can stay safe. An ______________ is a body part or a way an animal acts to stay alive. Giraffes have long necks to eat leaves from the tops of trees.

Some animals can ______________ into their environment. The color or ______________ of an animal can help it hide from other animals. This is called ______________ . The ______________ of spots on a leopard helps it hide. Some animals can grow fur and feathers of a different ______________ . A ptarmigan has brown feathers in the summer, but in the ______________ it will turn white. This helps it hide in the snow. Some animals travel in large ______________ . This prevents them from getting eaten.

© Macmillan/McGraw-Hill

Name ______________________ Date ____________

Helpful Traits

Write About It

Describe one of the animals in your book.
Where does it live? What do you think it eats?
What traits help it live in its environment?

Getting Ideas

Write the name of the animal you chose in the center circle. In the outer ovals, write details about the animal.

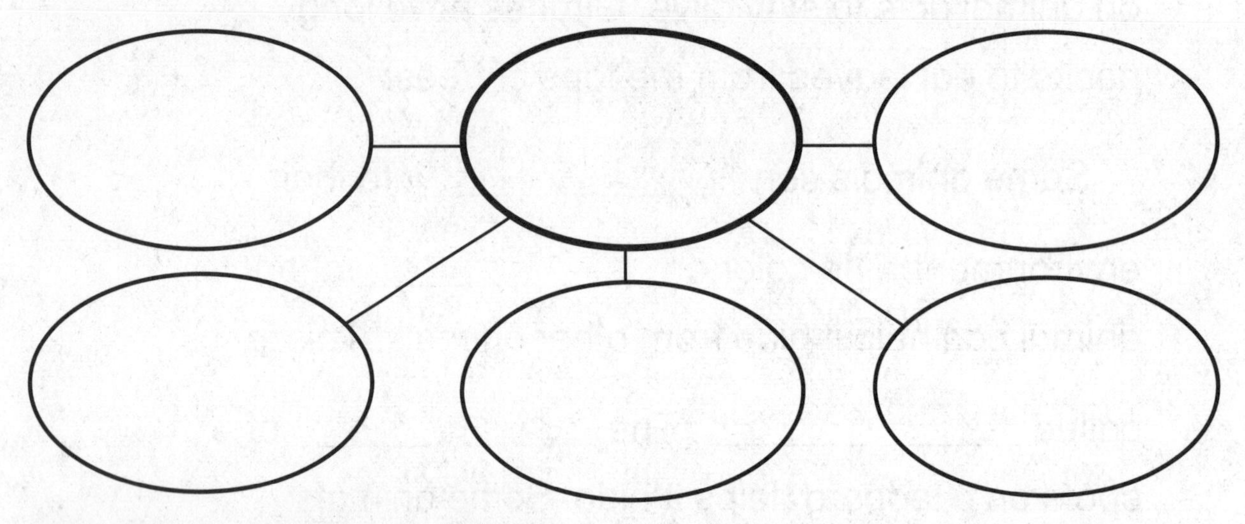

Planning and Organizing

Clifton wrote three sentences about jackrabbits. Write Yes if the sentence describes them. Write No if it does not describe them.

1. __________ They flatten their ears when they rest.
2. __________ Some have white fur in the winter.
3. __________ Jackrabbits have long tails.

© Macmillan/McGraw-Hill

Name ______________________ Date ____________

Places to Live

Fill in the blanks. Use the words from the box.

habitat	shelter	tunnels
plants	sunlight	

Where can plants and animals live? Living things can live in any ____________ where they get what they need to survive. Plants need soil, nutrients, water, and ____________ from their habitats in order to grow. Animals need food, water, and ____________ from their habitats in order to grow.

Plants and animals use their habitats in different ways. Some animals eat the ____________ and animals that live in their habitats. Other animals dig ____________ in the soil to hide from animals that want to eat them. Some plants even eat animals that live in their habitats!

© Macmillan/McGraw-Hill

LESSON
Outline

Name ______________________ Date __________

Food Chains and Food Webs

Use your book to help you fill in the blanks.

What is a food chain?

1. A ______________ shows how food energy moves from one living thing to another.

2. The ______________ is at the beginning of most food chains.

3. Plants need sunlight in order to grow, and ______________ eat plants in order to live.

4. Some food chains involve animals that live in the ______________, while others involve animals that live on land.

5. Some animals eat ______________ and animals that are no longer living.

6. Animals such as ______________ break up dead things into smaller pieces.

7. A ______________ is an animal that hunts and eats other animals.

8. Animals that are hunted by other animals are called ______________ .

© Macmillan/McGraw-Hill

Name ______________________________ Date ____________

LESSON
Cloze Activity

Food Chains and Food Webs

Fill in the blanks. Use the words from the box.

break	food web	predator	study
food chain	plants	prey	Sun

Different living things need different kinds of food in order to survive. A ______________ shows what an animal eats and where its food comes from. Scientists ______________ food chains to learn more about living things in our world.

Most food chains start with the ______________ . Plants use light and heat from the Sun to grow, then animals eat the plants. A ______________ is an animal that eats other animals. An animal that is hunted by a predator is called ______________ . Some living things eat dead ______________ and animals. They ______________ down the dead parts into pieces that become part of the soil. One kind of animal can be food for many animals. A ______________ shows how different food chains are connected. You are part of a food web too!

© Macmillan/McGraw-Hill

Name ______________________________ Date ____________

A Food Web for Lunch

Write About It

Explain how Emma, the chicken, the lettuce, and the wheat form a food web. Think about the food chains in Emma's lunch to help you form a food web of your own lunch.

Getting Ideas

Create a food web for your lunch.

Planning and Organizing

Put the steps in the correct order.

____________ Emma drinks milk for breakfast.

____________ The cow eats grass.

____________ A farmer milks the cow.

© Macmillan/McGraw-Hill

Name ______________________________ Date ____________

Drafting

Write a sentence to explain the food web. Tell your main idea.

____

Now write how the foods in Emma's breakfast form a food web. Start with the sentence you wrote above. Explain how the foods are connected.

Revising and Proofreading

Zack wrote some sentences. He made five mistakes. Find the mistakes. Then correct them.

The Son is the most important part of the food web. It gives energie to plants. The plants is eaten by the animals. Some animals then produce food. Chickens lay eggs. cows produce milk. Farmers gather the eggs for people to eat. Farmers also milk cows and bottle the milk. People drink the milk

Now revise and proofread your writing. Ask yourself:

- ▶ Did I explain the food web in Emma's breakfast?
- ▶ Did I tell the steps in order?
- ▶ Did I correct all mistakes?

© Macmillan/McGraw-Hill

Name ______________________ Date __________

Habitats Change

Use your book to help you fill in the blanks.

How do habitats change?

1. Habitats ______________ in many different ways.
2. A ______________ is one way nature can change a habitat.
3. A drought is a slow change that takes place when an area gets little or no ______________ for a long time.
4. Animals and ______________ can change habitats.

What happens when habitats change?

5. When habitats change, the ______________ and animals that live there may adapt or make changes.
6. Other plants and animals may not be able to find ______________ they need and can become endangered.
7. An animal becomes ______________ when many of its same kind die.

© Macmillan/McGraw-Hill

How can we tell what a habitat used to be like?

8. Scientists study ______________ to learn what Earth was like long ago.

9. Fossils can tell scientists how ______________, plants, and animals have changed over time.

10. Some fossils do not ______________ the habitat where they were found.

11. That tells scientists that there has been a ______________ in the habitat.

12. When an animal becomes ______________, there are no more of its kind left in the world.

Critical Thinking

13. Scientists have found fossils with fins and tails in dry areas. What do you think these places might have looked like long ago? How did they change?

____

____

____

____

____

© Macmillan/McGraw-Hill

Name ______________________ Date ____________

Habitats Change

Use the picture to answer the questions. Use the words in the box in your sentences.

drought	endangered	extinct	fossil

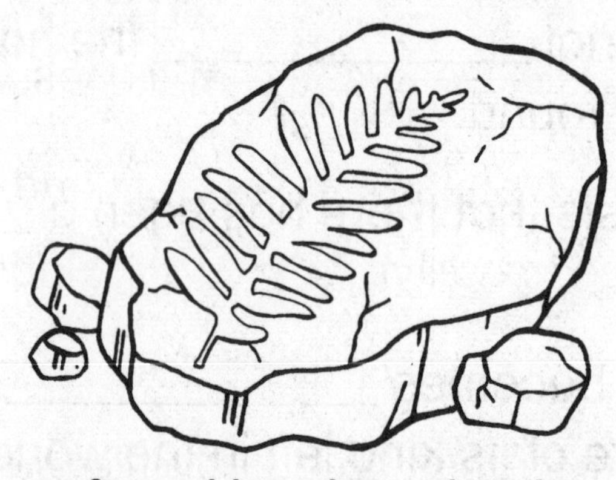

1. This fossil was found in a hot, dry desert. How do you think this habitat has changed over time?

____

____

____

____

2. How do you think this habitat became a desert?

____

____

____

© Macmillan/McGraw-Hill

Name ______________________ Date __________

LESSON
Cloze Activity

Habitats Change

Fill in the blanks. Use the words from the box.

~~change~~	~~endangered~~	fossil	people
~~drought~~	~~extinct~~	~~habitat~~	

Plants and animals live in different places. A habitat is a place where plants and animals live. People also live in habitats. Habitats can change over time. A drought changes a habitat when an area gets little or no rain for a long time. Habitats can change because of extinct, too. People destroy plant and animal homes by building roads and buildings.

When habitats change, plants and animals may die. A plant or animal becomes endangered when there are only a few of its kind left in the world. A plant or animal becomes extinct when there are no more of its kind left. When plants or animals disappear, they may leave a fossil behind. Scientists study fossils to learn what Earth was like long ago.

Fossil

© Macmillan/McGraw-Hill

Name ______________________ Date __________

Meet Mike Novacek

Read the Reading in Science pages in your book. As you read, think about how Mike and his team classify and categorize the fossils they discover. Mike has collected fossils of reptiles, mammals, and dinosaurs.

Use the chart below to classify the animals you have learned about. Remember, when you classify and categorize, you compare things. Then you put the ones that are alike into groups.

Fossils

Reptile	Mammal	Dinosaur

1. Where did you put the fossil of the Kryptobaatar skull in the chart?

____

© Macmillan/McGraw-Hill

Name ____________________ Date __________

Write About It

1. Classify and categorize. How can you put fossils into groups?

____

____

____

2. Why do you think scientists travel around the world looking for fossils?

____

____

3. What do you think a Kryptobaatar looked like? Draw a picture.

© Macmillan/McGraw-Hill

Name ______________________ Date ____________

Looking at Habitats

Fill in the blanks. Use the words in the box.

drought	extinct	predator
endangered	food chain	prey

1. An animal that hunts and eats another animal is called a ____________ .

2. An animal that is eaten by another animal is called ____________ .

3. A ____________ shows what an animal eats and where it gets its food.

4. An animal becomes ____________ when there are only a few of its kind left on Earth.

5. When an animal becomes ____________ , there are no more of its kind living on Earth.

6. A ____________ happens when a place gets little or no rain for a long time.

© Macmillan/McGraw-Hill

Name ____________________ Date __________

Draw pictures to complete the food chain.

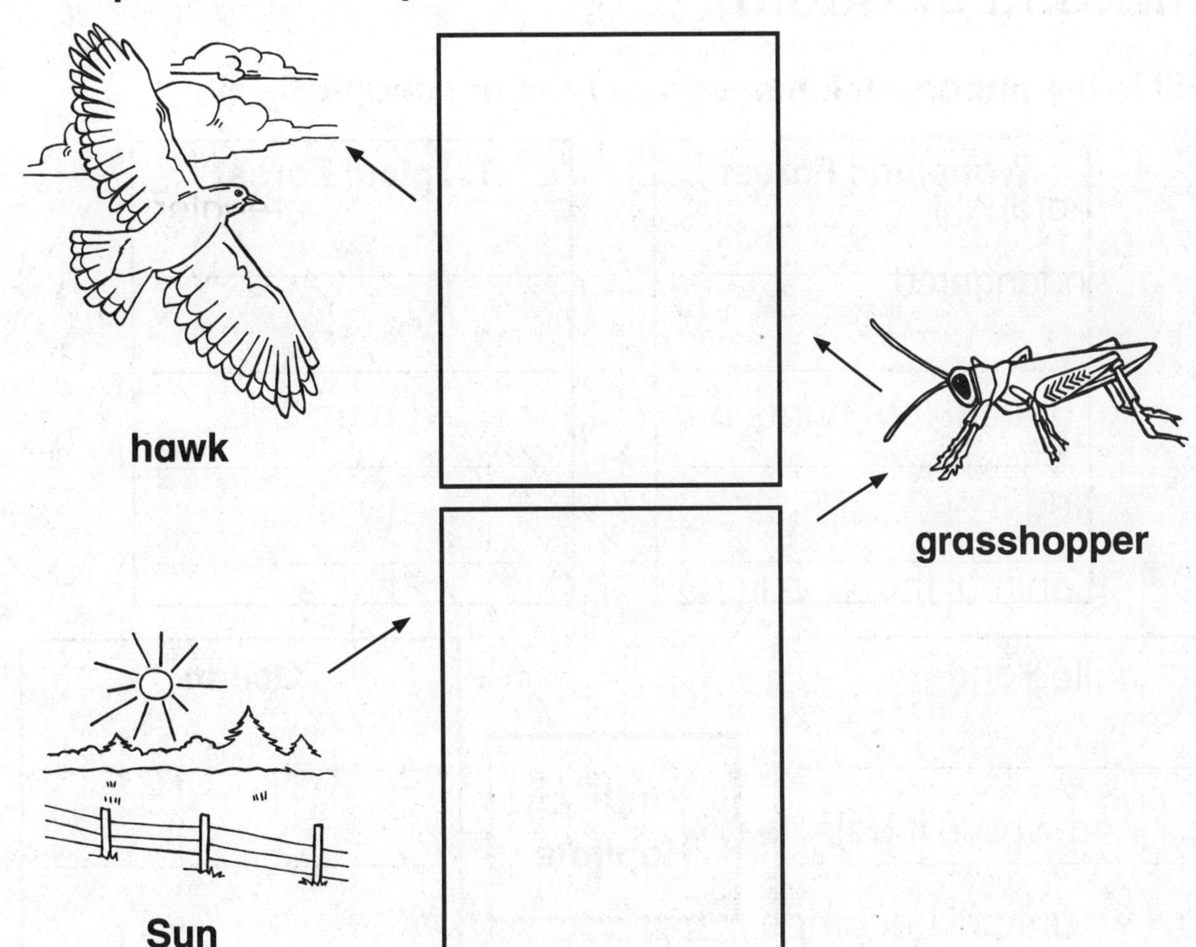

1. What is at the beginning of this food chain?

____

2. Is the animal that comes after the grasshopper a kind of predator or a kind of prey? Explain.

____

____

____

© Macmillan/McGraw-Hill

Name ______________________ Date __________

Kinds of Habitats

Fill in the important ideas as you read the chapter.

Kinds of Habitats

Woodland Forest

Tropical Forest

Pond

Ocean

Arctic

Desert

© Macmillan/McGraw-Hill

Name ______________________________ Date ___________

LESSON
Outline

Forests

Use your book to help you fill in the blanks.

What is a woodland forest like?

1. A ________________ habitat has many trees.
2. It is warm in the summer and ______________ in the winter.
3. A habitat is a place where ______________ and animals get what they need to live.
4. Most ______________ in the forest have leaves that change color in the fall.
5. Some trees have leaves that stay ______________ all year.
6. Animals can ______________ in a woodland forest in many ways.
7. Some animals eat leaves, ______________ , and nuts.
8. Other animals build homes in trees and ______________ in logs during the winter.

© Macmillan/McGraw-Hill

Name _______________ Date _______

What is a tropical rain forest?

9. A _______________ rain forest is a warm, steamy, moist place with many trees.

10. Some animals, such as birds, bats, and insects, live high in the _______________ .

11. Other animals such as jaguars, tapirs, and wild boars live on the _______________ .

12. Many trees grow very tall, have large _______________ , and block sunlight from falling to the ground below.

Critical Thinking

13. Why do you think animals in the tropical rain forest do not sleep all winter?

____

____

____

____

____

© Macmillan/McGraw-Hill

Name ______________________________ Date ____________

Forests

How do woodland forests and tropical rain forests compare? Fill in the Venn diagram.

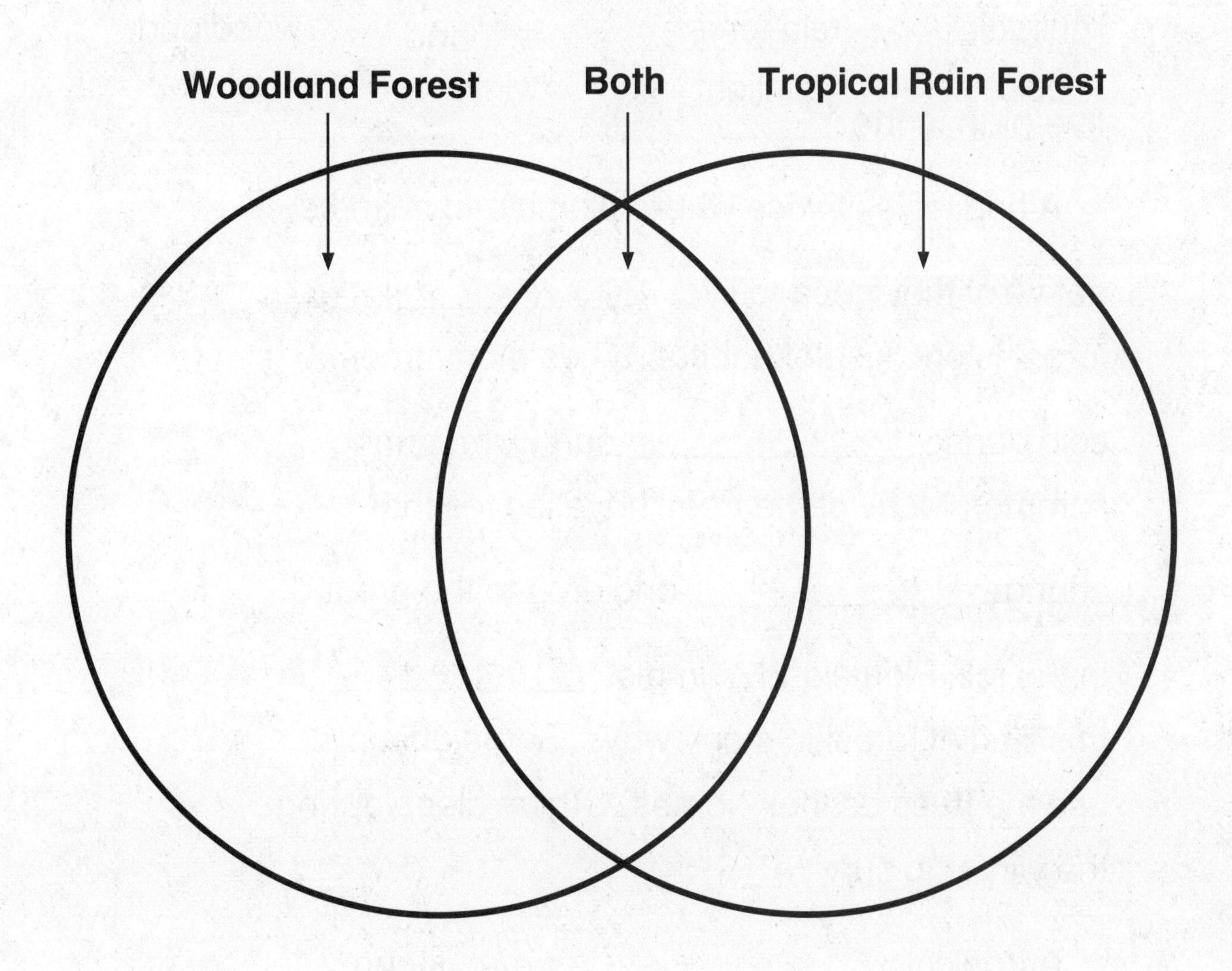

© Macmillan/McGraw-Hill

LESSON
Cloze Activity

Name ______________________________ Date ____________

Forests

Fill in the blanks. Use the words from the box.

animals	rain forest	survive	woodland
color	sunlight	winter	

A habitat is a place where plants and animals get what they need to live. A ______________ forest is one kind of habitat. It has many trees. It is cold during ______________ and warm during summer. Many of the trees have leaves that change ______________ and drop to the ground in the fall. Plants and animals ______________ in this kind of forest in many ways. Some animals use the trees as their homes. Others sleep during the winter to survive.

A tropical ______________ is warm, steamy, and moist. The trees are tall and have very large leaves. They block ______________ from getting to the ground. Some ______________ live in the treetops. Other animals live on the ground.

© Macmillan/McGraw-Hill

Meet Liliana Dávalos

Read the Reading in Science pages in your book. As you read, think about how Liliana compares and contrasts things in her work as a biologist at the American Museum of Natural History. Remember, when you compare things, you decide how they are alike. To contrast is to decide how things are different.

Answer the questions and fill in the chart below.

1. What other habitats have you learned about in this chapter?

____

2. How is the rain forest alike and different from other kinds of forests?

Rain Forest	Regular Forest	Both

© Macmillan/McGraw-Hill

Name ______________________________ Date ____________

Write About It

1. Compare and Contrast. How would life change for the manakins if the Amazon rain forest were cut down? Would it be the same as it is today? Explain.

____

____

____

____

2. A biologist is a scientist who studies living creatures. What other kinds of scientists have you learned about? How are they alike and different?

____

____

____

3. Biologists, like Liliana, often compare and contrast animals in their work. Why?

____

____

© Macmillan/McGraw-Hill

Name ______________________ Date ____________

Hot and Cold Deserts

Use your book to help you fill in the blanks.

What is a hot desert like?

1. A ______________ is a very dry and sandy habitat.

2. This kind of habitat can be ______________ during the day and cool at night.

3. It does not ______________ often in the desert.

4. Plants in this habitat survive by storing ______________ in their stems and leaves.

5. Some desert plants have ______________ that spread far out from the plant.

6. Desert animals get water from eating ______________ or other animals.

7. Most desert animals sleep during the day and hunt for ______________ at night.

© Macmillan/McGraw-Hill

Name ______________________________ Date ____________

What is the Arctic like?

8. The ______________ is a very cold and windy desert near the North Pole.

9. Many animals that live in this habitat have thick ______________ that keeps them warm.

10. Other animals have a thick layer of fat, called ______________, to keep warm.

11. Plants in the Arctic grow close to the ground to stay safe from the cold ______________.

Critical Thinking

12. Do you think that plants in hot and cold desert habitats store water in the same way? Why or why not?

____

____

____

____

____

____

© Macmillan/McGraw-Hill

Name ______________________ Date ____________

LESSON
Outline

Oceans and Ponds

Use your book to help you fill in the blanks.

What is the ocean like?

1. The largest bodies of water on Earth are called ______________.
2. An ocean is a large body of ______________ water.
3. Most of ______________ is covered by oceans.
4. Kelp is a kind of ______________, or ocean plant.
5. It grows in the ocean and provides ______________ for many ocean animals.
6. Animals in the ocean have ______________ parts that help them swim through the water.
7. Some animals in the ocean have ______________, spines, or stingers to help them stay safe.

© Macmillan/McGraw-Hill

Name ______________________ Date ____________

What is a pond like?

8. A ______________ is much smaller than an ocean.

9. Ponds have ______________ water and do not flow.

10. Frogs, fish, and ______________ are some animals that live in or near ponds.

11. Many plants grow in ______________ pond water near the shore.

12. Animals that live in ponds ______________ in different ways.

Critical Thinking

13. Do you think that the same types of animals live in both oceans and ponds?

____

____

____

____

____

____

© Macmillan/McGraw-Hill

Name ______________________ Date __________

Oceans and Ponds

Look at the animal and plant pictures beneath the box. Write the name of each animal or plant under the habitat where they live.

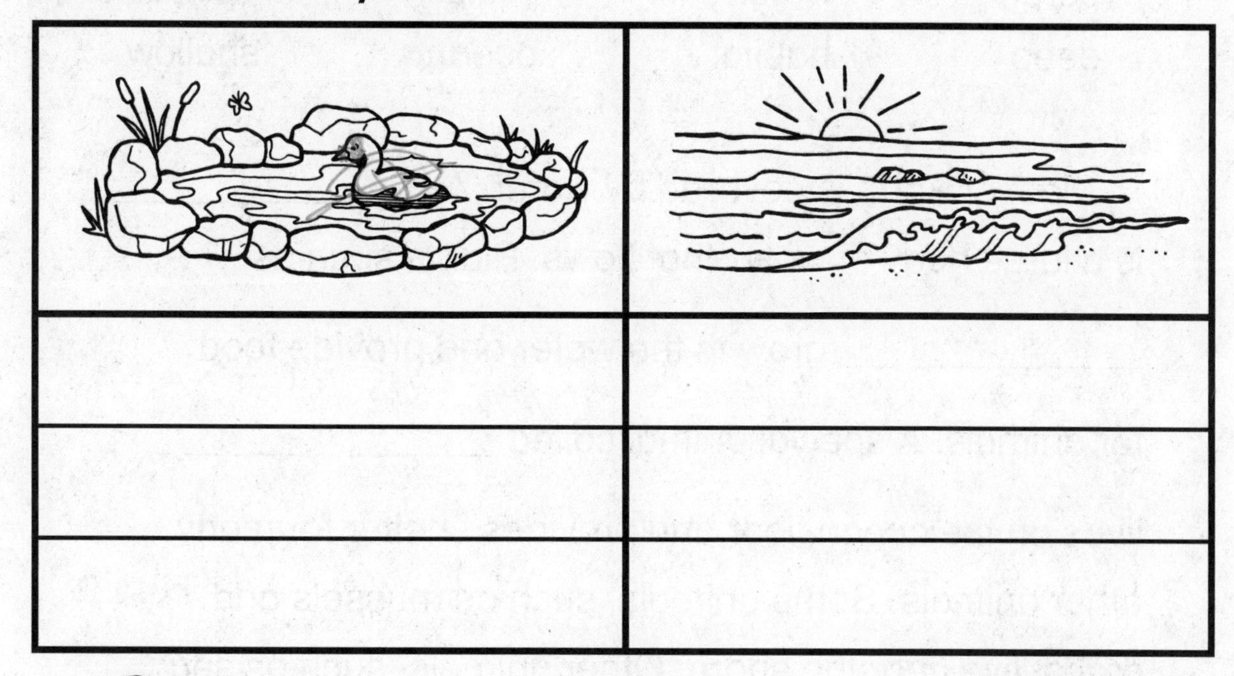

salamander

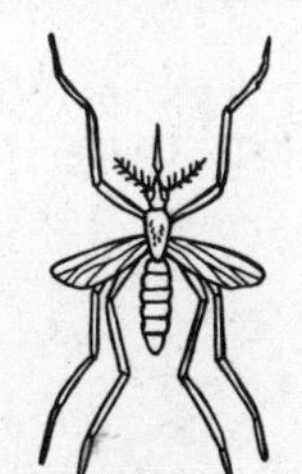

mosquito

dolphin

cat tails

penguin

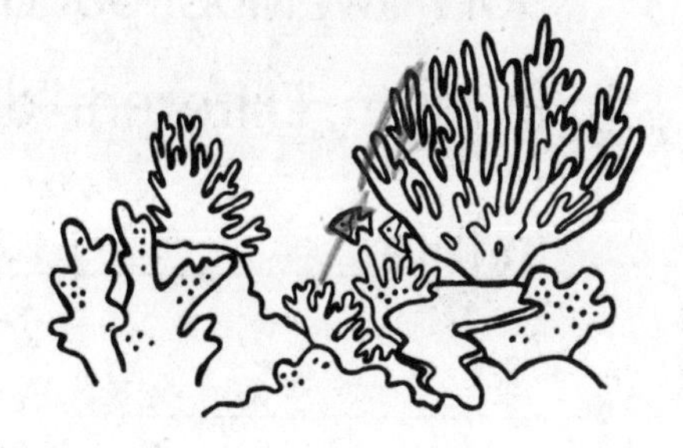

coral reef

© Macmillan/McGraw-Hill

LESSON
Cloze Activity

Name ______________________ Date __________

Oceans and Ponds

Fill in the blanks. Use the words from the box.

coral	fresh	kelp	pond
deep	habitat	ocean	shallow

Most of Earth is covered by water. An ______________ is a large body of water that flows. Plants such as ______________ grow in the water and provide food for animals. A special animal called ______________ lives on the ocean floor and provides shelter for many other animals. Some animals, such as mussels and crabs, live near the shore. Other animals, such as sea cucumbers and sea spiders, live in ______________ waters.

A ______________ is a body of water that does not flow. Most ponds have ______________ water in them. Different kinds of plants and animals live in this ______________ . Some plants grow in ______________ water near the shore. Their stems and leaves rise to the top of the water.

© Macmillan/McGraw-Hill

Name ______________________________ Date ___________

A Visit to the Ocean

Write a story about a trip you might take to the ocean. How would you get there? Who would you go with? Describe in your story what you would see, hear, and do. Write how it might feel to be there.

Getting Ideas

Picture yourself standing on a beach next to the ocean. Write what you see and hear.

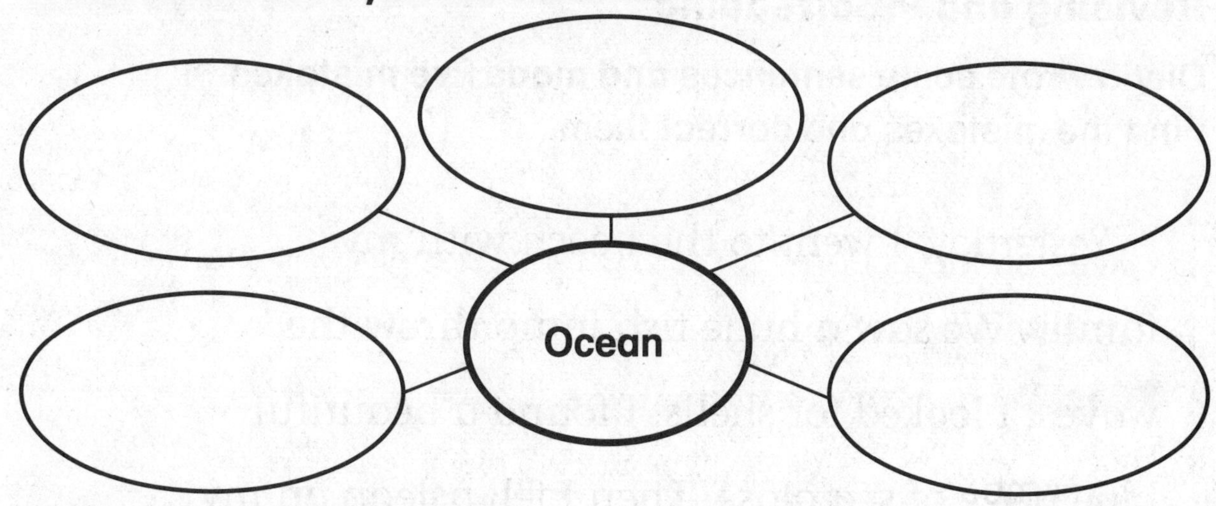

Planning and Organizing

Jackson wrote three sentences. They describe what he saw, heard, and did at the ocean. Circle the descriptive words he used.

1. The gigantic ocean waves roared loudly.
2. I saw white gulls sitting on a big rock near the shore.
3. I found a piece of green sea glass and two pretty pink shells.

© Macmillan/McGraw-Hill

Name ______________________________ Date ____________

Drafting

Write a sentence to begin your story. Use I to tell about yourself. Tell where you went and when.

____

Now write a story on a separate piece of paper. Put the events in time order. Describe what you saw, heard, and did at the ocean.

Revising and Proofreading

Olivia wrote some sentences and made five mistakes. Find the mistakes and correct them.

Yesterday, I went to the beech with my family. We saw a huge fish jump threw the waves. I looked for shells. I found a beautiful blue peice of sea glass. Then I fell asleap on my beach towel. When I wake up, it was almost time to go home.

Now revise and proofread your writing. Ask yourself:

- Did I tell how I got to the ocean and with whom I went?
- Did I describe what I saw, heard, and did?
- Did I correct all mistakes?

© Macmillan/McGraw-Hill

Name ______________________ Date __________

Earthworms

Soil Helpers

Read the Unit Literature pages in your book.

Write About It

Response to Literature

1. What do you think would happen to soil if there were no earthworms?

____

____

____

2. Can you imagine what the world looks like to an earthworm? Use the article to give you ideas. Draw a picture.

© Macmillan/McGraw-Hill

Name ______________________ Date __________

Land and Water

Fill in the important ideas as you read the chapter.
Use the words in the box.

continent	lake	ocean	stream
earthquake	landslide	plain	valley
flood	mountain	pond	volcano

What do you know about the Earth's land and water?

What is land like on Earth?

____

____

____

____

What is water like on Earth?

____

____

____

____

How can Earth's land and water change?

____

____

____

____

© Macmillan/McGraw-Hill

Name ______________________ Date ____________

LESSON
Outline

Earth's Land

Use your book to help you fill in the blanks.

What is land like on Earth?

1. Earth's land is smooth, ____________, or flat in many places.

2. It also has many ____________, or land shapes.

3. A ____________ is a high and rocky landform.

4. A ____________ is a low and flat landform.

What can maps tell us about Earth?

5. A ____________ shows where land and water are on Earth.

6. A ____________ is a map of Earth in the shape of a ball.

7. Maps can show where land is high or low on a

____________ or an island.

© Macmillan/McGraw-Hill

Name ______________________ Date __________

What is inside Earth?

8. There are three main ______________ inside Earth.

9. We live on the ______________, or outer layer of Earth.

10. The ______________ is the very hot layer below the crust.

11. The ______________ is the third layer of Earth.

12. It is part ______________, part liquid, and is very hot!

Critical Thinking

13. Why do you think plants and animals only live on Earth's crust?

____

____

____

____

____

© Macmillan/McGraw-Hill

Name ______________________________ Date ____________

Earth's Land

Follow the directions to color the map.

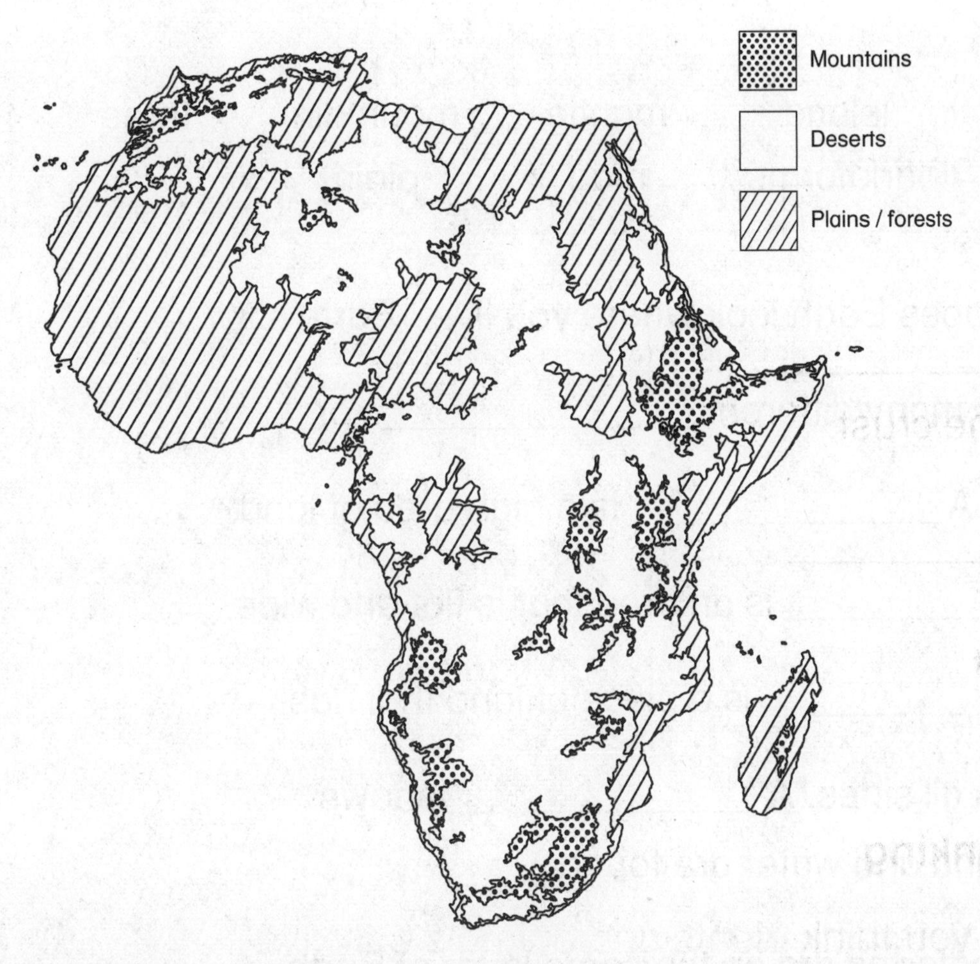

1. Outline the continent in black.

2. Color the mountains brown.

3. Color the deserts pink.

4. Color the plains and forests yellow.

5. Outline the islands in blue.

© Macmillan/McGraw-Hill

Name ______________________ Date ____________

Earth's Land

Fill in the blanks. Use the words from the box.

core	island	mantle	mountain	three
crust	landforms	map	plain	

How does Earth look where you live? Earth's land has many different ______________ , or shapes. A ______________ is a high area of land. A ______________ is an area that is flat and wide. An ______________ is a piece of land that has water on all sides. A ______________ shows where land and water are found.

All landforms are on the same layer of Earth. There are ______________ main layers that make up Earth. We live on the ______________ , Earth's top layer. The ______________ is the very hot middle layer. The center of Earth is called the ______________ . It has solid and liquid parts. It is even hotter than the mantle!

© Macmillan/McGraw-Hill

Name ______________________ Date __________

Earth's Water

Use your book to help you fill in the blanks.

Why is Earth's water important?

1. All living things on Earth need ____________ to survive.
2. People and animals drink water and also use it to take ____________ .
3. Plants use water to ____________ food and carry nutrients to every plant part.
4. Most living things can only drink ____________ water.
5. Lakes, ponds, ____________ , and streams all have fresh water.
6. Water comes from ____________ and snow that melt and flow down hills and mountains.

Where is most of Earth's water found?

7. The ____________ surround every island and continent.

© Macmillan/McGraw-Hill

Name ______________________ Date ____________

8. Oceans are large, deep bodies of ______________ water.

9. People can not ______________ ocean water.

10. Most of ______________ is covered by oceans.

11. Many ______________ and animals live in the ocean.

12. People build ______________ to carry goods and people around the world.

Critical Thinking

13. What kind of water did you use today? Was it fresh water or ocean water? How did you use it?

____

____

____

____

© Macmillan/McGraw-Hill

Name ______________________________ Date ____________

Earth's Water

Fill in the missing letters to complete each sentence.

1. A large body of salty water is called an
   _ c _ _ n.

2. A _ o _ _ is a small body of water that has land on all sides.

3. People clean f _ e _ _ water so they can use it.

4. First, water is brought from lakes through
   _ i _ e _.

5. Then, special machines _ _ e _ n the water.

6. Finally, water goes through pipes to reach our
   h _ _ e _.

© Macmillan/McGraw-Hill

Name ____________________ Date __________

Earth's Water

Fill in the blanks. Use the words from the box.

animals	lakes	ponds	three-fourths
ice	ocean	salty	water

People, plants, and animals all over the world use water every day. Most living things need fresh ______________ to survive. It is found in lakes, ______________, rivers, and streams. When ______________ and snow melt, water flows into streams and rivers. People can clean this water to use for drinking, cooking, cleaning, and playing.

But most of our world's water is not in ______________, ponds, rivers, or streams. Oceans cover ______________ of Earth. An ______________ is a large, deep, salty body of water. Many plants and ______________ live in the ocean. Ocean water is too ______________ for people to drink. It's just right for swimming and sailing!

© Macmillan/McGraw-Hill

Name ______________________________ Date ____________

My Water

Write About It

Write a report about lakes, streams, or ponds where you live. Tell what animals live there and how you can help protect them and the water. Draw a picture of the water. Share your report with the class.

Getting Ideas

Choose a body of water where you live. Write it in the center circle. Write which animals live there.

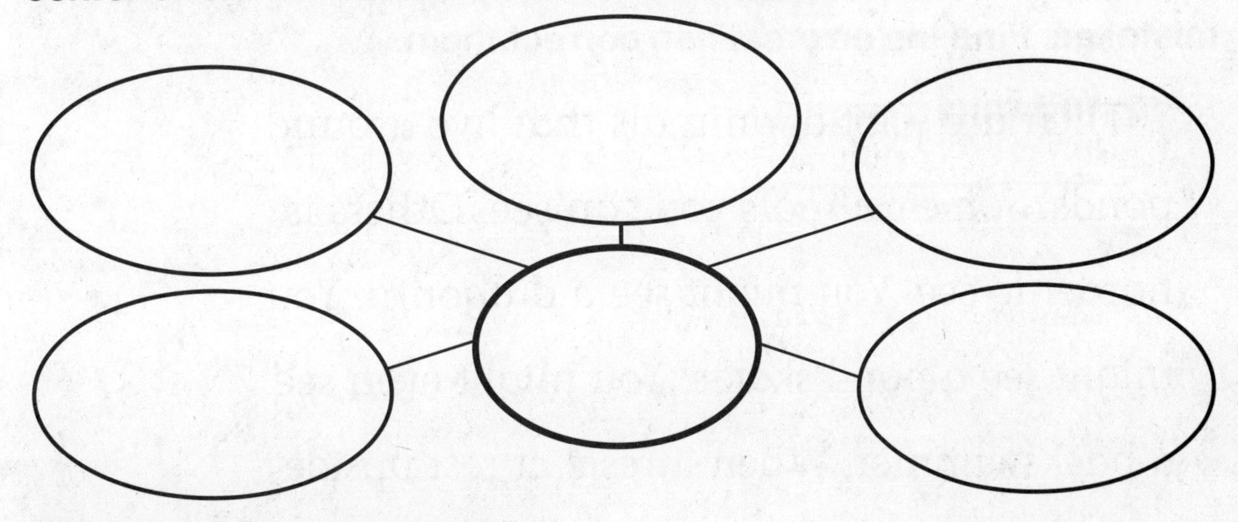

Planning and Organizing

Mai Ling wrote about pond life. Write Yes if the sentence is true. Write No if it is not true.

1. ______________ Frogs and minnows live in ponds.
2. ______________ Ducks live under the water.
3. ______________ Raccoons use ponds to find fish to eat.

© Macmillan/McGraw-Hill

Name ______________________________ Date ____________

Drafting

A report gives facts about a topic. A topic sentence tells the main idea. Start your report by writing a topic sentence. Name the body of water you are writing about.

____

Now write your report on a separate piece of paper. Tell about animals that live in the lakes, streams, or ponds where you live. Tell how you can protect them.

Revising and Proofreading

Mai Ling wrote some sentences. She made five mistakes. Find the errors. Then correct them.

Their are a lot of animals that live around ponds. Some animals you can see. Others is harder to see. You might see a dragonfli. You might see a pond skater? You might even see a backswimmer. When it rests, it rest upside down on the top of a pond.

Now revise and proofread your writing. Ask yourself:

- Did I write a topic sentence?
- Did I tell what animals live in the body of water?
- Did I correct all mistakes?

© Macmillan/McGraw-Hill

Name ______________________ Date ____________

LESSON
Outline

Changes on Earth

Use your book to help you fill in the blanks.

How does Earth change slowly?

1. It takes a long time for Earth's land to ____________.

2. Fast-moving ____________ can change rocks and mountains.

3. Wind blows ____________ and carries away soil.

4. The ____________ of the land or rock changes shape.

5. Ice can also change ____________ over time.

6. Water can get inside ____________ in rocks.

7. The water ____________ into ice and breaks the rocks apart.

© Macmillan/McGraw-Hill

Name ______________________________ Date ___________

How does Earth change quickly?

8. Sometimes ______________ changes quickly.

9. An ______________ changes land when Earth's crust shakes.

10. A ______________ can also change the shape of land.

11. A ______________ is another fast change that happens when a lot of rain falls quickly.

12. Rocks and ______________ can move from high to low ground during a landslide.

Critical Thinking

13. What can make Earth change?

____

____

____

____

____

© Macmillan/McGraw-Hill

Name ______________________________ Date ____________

Living With Floods

Read the Reading in Science pages in your book. Read each paragraph. Then make an inference based on the statement in the "What I Know" column. Write your inference in the chart.

What I Know	What I Infer
The title of the article is "Living with Floods."	
Each year, heavy rains cause the Mekong River to flood between May and November.	
Rice can not grow in salty water.	

© Macmillan/McGraw-Hill

Name ______________________ Date __________

1. What did you learn about the dry season in Vietnam?

2. Compare how the Mekong River looks during the wet season and the dry season.

Write About It

How do people in this part of the world use water from floods and the sea to help them live?

© Macmillan/McGraw-Hill

Name ______________________ Date ____________

Land and Water

Read the word in each box.

Color the box blue if it tells how Earth can change slowly.

Color the box green if the word tells how Earth can change quickly.

Color the box yellow if the word tells what is inside Earth.

Color the box red if the word tells about a landform.

Some boxes will not be colored at all.

continent	water	core	ice
earthquake	hill	flood	wind
map	mountain	animals	plain
landslide	mantle	crust	valley

© Macmillan/McGraw-Hill

Name ______________________ Date __________

Draw a line from the word to its meaning.

1. earthquake	**a.** the center of Earth
2. core	**b.** a large body of salty water
3. island	**c.** an opening in Earth's crust
4. ocean	**d.** when Earth's crust shakes
5. landslide	**e.** a very large piece of land
6. volcano	**f.** a small piece of land that has water on all sides
7. continent	**g.** when rocks move from higher to lower ground

© Macmillan/McGraw-Hill

Name ______________________ Date ____________

Earth's Resources

Fill in the important ideas as you read the chapter. Write at least one way we use each of the natural resources shown on the left. Then, answer the question.

How do we use Earth's resources?

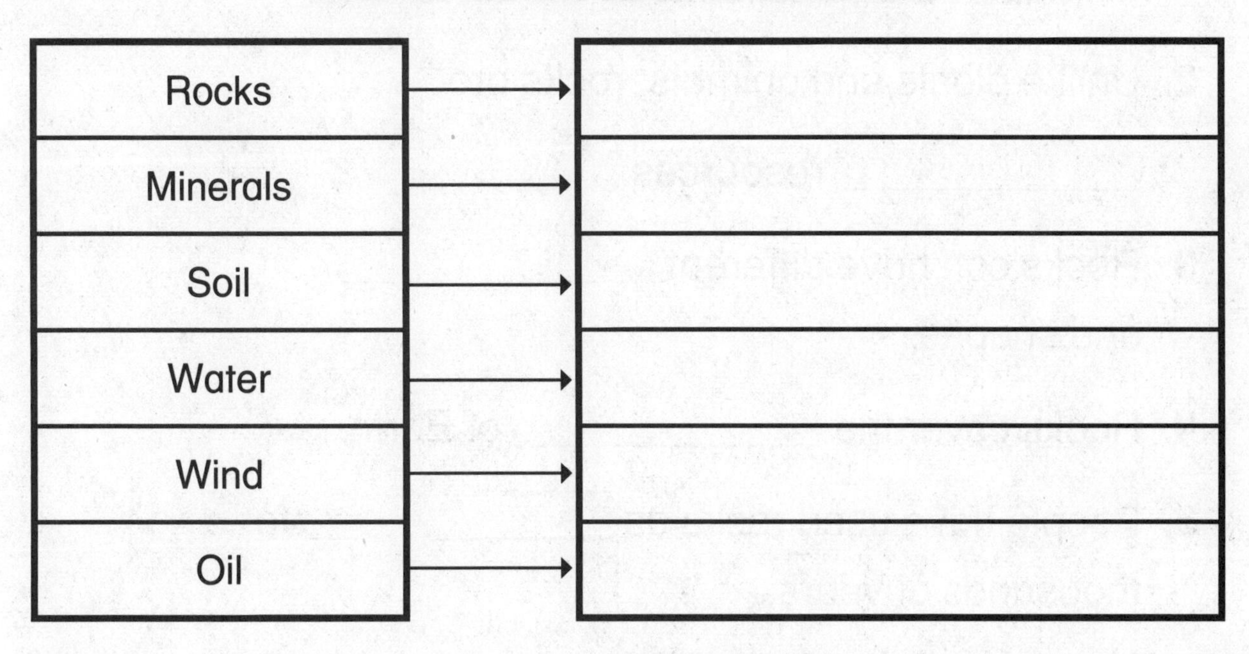

Why should we care for Earth's resources?

____

____

____

____

© Macmillan/McGraw-Hill

Name ______________________________ Date ____________

Rocks and Minerals

Use your book to help you fill in the blanks.

What are rocks?

1. We use ________________ like plants, animals, water, and rocks every day.
2. Unlike plants and animals, rocks are ______________ resources.
3. Rocks can have different ______________ and shapes.
4. Rocks cover the ______________ of Earth.
5. People have used rocks as ______________ for thousands of years.
6. People can also use rocks to carve ______________ or build things.

© Macmillan/McGraw-Hill

Name ______________________ Date __________

What are minerals?

7. All ______________ are made of one or more minerals.

8. A ______________ is a nonliving thing that comes from Earth.

9. It takes ______________ of years for rocks and minerals to form inside Earth.

10. People must ______________ to find rocks and minerals.

11. People use minerals like ______________ to help make toothpaste, steel, and other materials.

Critical Thinking

12. Why are rocks and minerals natural resources?

____

____

____

____

____

© Macmillan/McGraw-Hill

LESSON
Vocabulary

Name ______________________ Date __________

Rocks and Minerals

Fill in the blanks. Then find the vocabulary words in the puzzle.

1. A ______________ is a hard, nonliving part of Earth.

2. A rock can be made of one ______________ or made of many different kinds.

3. A ______________ resource is something from nature that people use.

4. People can make ______________ out of rocks.

5. The mineral ______________ can be found in a pencil.

N	A	T	U	R	A	L	S	T	T
R	M	I	N	E	R	A	L	F	O
O	K	P	M	T	R	B	I	U	O
C	F	G	A	E	T	O	M	S	L
K	G	R	A	P	H	I	T	E	S

© Macmillan/McGraw-Hill

Name ______________________ Date __________

LESSON
Cloze Activity

Rocks and Minerals

Fill in the blanks. Use the words from the box.

graphite	natural resource	tools
magnetite	statues	
minerals	surface	

Rocks are the most common materials on Earth. They cover the ______________ of Earth, from the top of a mountain to the bottom of the ocean. Rocks and ______________ are nonliving things that make up part of Earth's surface.

Rocks and minerals are natural resources. A ______________ is something from nature, such as water, wood, or minerals, that people use in everyday life. The mineral ______________ is found in magnets, and ______________ is found in pencils. For thousands of years, people have made ______________ from rocks. They have even made ______________ from rocks. The Sphinx in Egypt was carved from rock thousands of years ago.

© Macmillan/McGraw-Hill

Name ______________________ Date __________

Rock and Stroll

Write About It

Write a letter to a friend. Write about a walk you took. Describe the rocks you saw. Explain how you think they got their shape.

Getting Ideas

Fill in the chart. In the first column, tell what rocks you saw. In the second column, describe them.

Types of Rocks	Details

Planning and Organizing

Write Yes if the sentence describes a rock. Write No if it does not.

1. __________ The big boulder was gray and black.
2. __________ The small stones were smooth and oval.
3. __________ I really like to climb rocks at the beach.

© Macmillan/McGraw-Hill

Name ______________________ Date __________

Drafting

Write a greeting and first sentence for your letter. It should tell where you took your walk.

____

____

Now write your letter on a separate piece of paper. Describe the rocks you saw, and sign your name.

Revising and Proofreading

Fill in the blanks with words from the box.

gigantic	heavy	tall
gray	small	weird

This morning, Zoe and I walked in the park. We saw a ____________ rock. It was very ____________. The stone was a deep ____________. The rock had a ____________ shape. I think that the ____________ rain wore the rock down. The rocks by a pond were very ____________ and white.

Now revise and proofread your writing. Ask yourself:

- ▶ Did I describe the rocks and how they got their shapes?
- ▶ Did I correct all mistakes?

© Macmillan/McGraw-Hill

LESSON
Outline

Name ______________________________ Date ____________

Soil

Use your book to help you fill in the blanks.

What is soil?

1. Earth's ______________ is made of a mix of sand, clay, rocks, and minerals.
2. Parts of ______________ and animals that have died are in soil, too.
3. Clay soil, topsoil, and ______________ are found in different places and have different colors.
4. Each kind of soil feels different and has a different ______________ .
5. Some soils feel like ______________ or pebbles.
6. Other soils feel ______________ and are light in color.
7. Some soils hold more ______________ than others.
8. The soils that hold more water have a ______________ color.

© Macmillan/McGraw-Hill

Name ______________________________ Date ____________

LESSON
Outline

How is soil formed?

9. It can take a very long time for rocks and

______________ to break down into soil.

10. When plants and animals die, their parts

______________ and rot away.

11. The ______________ that were once inside living things make the soil healthy for plants.

12. Plants grow best in ______________ .

13. Topsoil is the ______________ of soil with decaying plant and animal parts.

14. A mix of soil and parts of rotting plants and

animals is called a ______________ pile.

Critical Thinking

15. Why is soil important?

____

____

____

____

© Macmillan/McGraw-Hill

Name ______________________ Date __________

Soil

Match each word in the box to the correct picture and use the word in a sentence.

compost	decompose	topsoil

1.

____

____

____

2.

____

____

____

3.

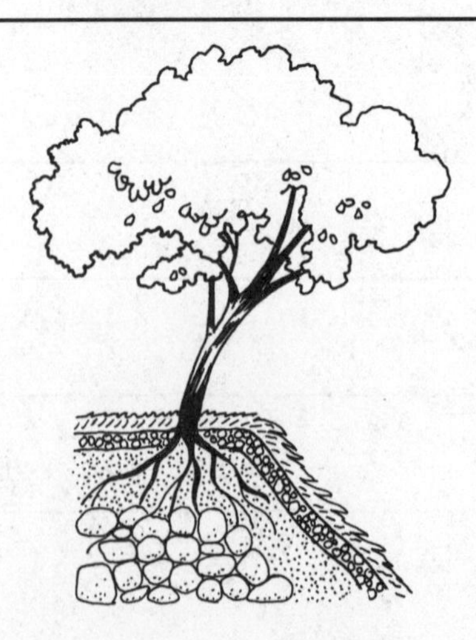

____

____

____

____

© Macmillan/McGraw-Hill

Name ______________________ Date ____________

LESSON
Cloze Activity

Soil

Fill in the blanks. Use the words from the box.

decompose	natural resources	rocks
layer	nutrients	texture

Soil can be found almost everywhere on land. Soil is one of Earth's most important ________________. Soil is formed when ______________ and minerals break down into smaller pieces over many years. Parts of dead plants and animals ______________ and become part of the soil, too. The ______________ inside these once-living things help make the soil healthy.

Plants grow best in the top ______________ of soil, called topsoil. This is where the soil is richest with nutrients. Some soils are light, and others are dark. Each soil feels different and has a different ______________. Some soils hold a lot of water, while others are sandy and do not hold much water. However, all soils are important to Earth.

© Macmillan/McGraw-Hill

LESSON
Outline

Name ______________________________ Date ____________

Using Earth's Resources

Use your book to help you fill in the blanks.

How do we use natural resources?

1. People use air, wind, water, rocks, and soil as ____________________ every day.

2. Earth can quickly ________________ resources such as water and wind.

3. Other resources, such as ________________, can not be made quickly by Earth.

Why should we care for Earth's resources?

4. It is important to care for Earth's ________________, water, and air.

5. Pollution can harm living and ________________ things such as plants, animals, and people.

6. Pollution makes Earth's air, ________________, and land dirty.

7. To stop land pollution, people can clean up the ________________ they leave behind.

© Macmillan/McGraw-Hill

Name ______________________________ Date ____________

LESSON **Outline**

How can we save Earth's resources?

8. People can help to ______________ Earth's resources.

9. Remember the ______________ Rs: reduce, reuse, and recycle.

10. When people ______________, they cut back on how much they use a resource.

11. When people ______________ something, they use it again, often in a new way.

12. When people ______________ glass, paper, and cans, they make new things out of them and reduce litter.

Critical Thinking

13. How do you use natural resources every day?

____

____

____

____

____

© Macmillan/McGraw-Hill

Name ______________________ Date ____________

Using Earth's Resources

Each picture below shows a way to conserve Earth's natural resources. Write reduce, reuse, or recycle under the correct picture.

1.

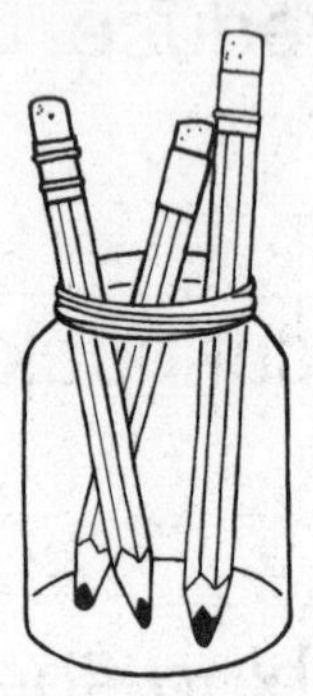

____

2.

____

3.

____

4.

____

What are other ways you can help conserve Earth's resources where you live?

____

____

____

© Macmillan/McGraw-Hill

Name ______________________________ Date ____________

LESSON
Cloze Activity

Using Earth's Resources

Fill in the blanks. Use the words from the box.

conserve	pollution	wind
litter	recycle	
natural resources	reduce	

Earth needs your help. Every day, you use ______________ such as air, water, and land. Earth can quickly replace resources like water and ______________ . Resources such as minerals take longer to replace. It is important to ______________ Earth's resources.

Something that makes air, water, or land dirty is called ______________ . Help keep land and water clean by picking up ______________ . You can protect resources if you ______________ and reuse things. You can ______________ paper, glass, and plastic so they can be made into something else. Remembering the 3 Rs is the first step to helping save Earth's resources.

© Macmillan/McGraw-Hill

Name ____________________ Date ____________

A World of Wool

Read the Reading in Science pages in your book. As you read, pay attention to the most important ideas. List them in the chart below. Then summarize the article. Remember, when you summarize, you retell the most important ideas in the selection.

Idea #1

____

____

____

Idea #2

____

____

____

Idea #3

____

____

____

Summary

____

____

____

____

____

____

____

____

____

____

____

© Macmillan/McGraw-Hill

Name ______________________ Date ___________

Write About It

Summarize. Write a paragraph that retells what you learned about llama wool. Use the following words in your writing: cold, warm, sweaters, llamas, camels, fur, spin, yarn, clothes, Andes Mountains.

____

____

____

____

____

____

____

____

____

____

____

____

____

© Macmillan/McGraw-Hill

Name ______________________ Date __________

Earth's Resources

Write a short story about what is happening in the picture. Use at least three words from the box.

conserve	natural resources	reduce
litter	pollution	reuse
minerals	recycle	

Title: ______________________

Story: ______________________

____

____

____

____

____

© Macmillan/McGraw-Hill

Name ______________________ Date __________

If the sentence is true, write TRUE. If the sentence is not true, write FALSE.

1. ______________ Rocks are made of minerals.
2. ______________ Litter is garbage that people leave behind.
3. ______________ Plastic is a natural resource.
4. ______________ When dead plants or animals decompose, their parts rot away.
5. ______________ Soil is made only of rocks.
6. ______________ A compost is a mix of paper, plastic, and glass.

© Macmillan/McGraw-Hill

Name ______________________ Date __________

Sunflakes

By Frank Asch

Read the Unit Literature pages in your book.

Write About It

Response to Literature

1. What season is the poet writing about? Use the poem to tell how you know.

____

____

2. What are some things that you do in July? How do your activities compare to the poet's?

____

____

3. What do you think a sunflake looks like? Draw a picture.

© Macmillan/McGraw-Hill

Name ______________________ Date __________

Weather

Fill in the blanks. Use the words from the box.

anemometer	rain gauge	weather
Fahrenheit	temperature	wind
precipitation	thermometer	wind sock

Look out the window. What is the ______________ like? Is it sunny? Is it rainy? People use special tools to find out about the weather. A ______________ is used to find out how hot or cold it is outside. This tool measures the ______________ of the air. Temperature is measured in degrees ______________ or in degrees Celsius.

Moving air is called ______________. The speed with which the wind blows is measured by using an ______________. A ______________ shows what direction the wind is blowing. Rain, snow, sleet, and hail are kinds of ______________.

A ______________ is used to measure precipitation. These tools help people learn about the weather.

© Macmillan/McGraw-Hill

Name ______________________ Date __________

A Snowy Day

Write About It

Write a story about what you might do on a snowy day.

Getting Ideas

Picture a snowy day in your mind. Now put yourself in the picture. Write what you are doing.

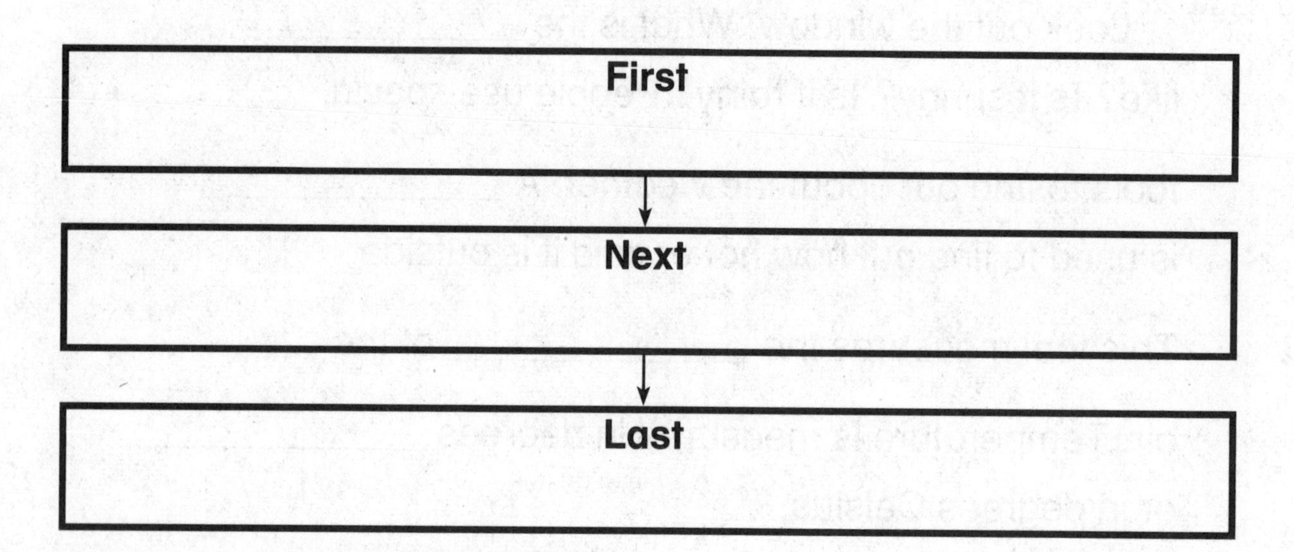

Planning and Organizing

Put the sentences in time order.

_________ We bundled up in warm clothing.

_________ We climbed to the top of the hill and slid down.

_________ We walked to the big hill in the park.

© Macmillan/McGraw-Hill

Name ______________________ Date ____________

Drafting

Write the first sentence of your story. Tell how you started your snowy day.

____

Now write your story on a separate piece of paper. Put the events in time order. Include details.

Revising and Proofreading

Use the words in the box to fill in the blanks.

cold	long	warm
huge	soft	

It was a cloudy and ____________ day. Andy and I wore ____________ clothes outside. We noticed ____________, narrow icicles hanging from the trees. They were beautiful! Maple Hill was covered in ____________, deep snow that made it hard to climb. At the top, we made a ____________ ball of snow. Then we rolled it down the hill.

Now revise and proofread your writing. Ask yourself:

- ▶ Did I use details to tell what I might do on a snowy day?
- ▶ Did I correct all mistakes?

© Macmillan/McGraw-Hill

Name ______________________ Date __________

The Water Cycle

Use your book to help you fill in the blanks.

How does water disappear?

1. Water ______________ when it gets very warm.
2. When water evaporates, it changes from a ______________ to water vapor.
3. Water vapor is in the form of a ______________ .
4. When water ______________ , it changes from a gas to a liquid.
5. When the air ______________ , the water vapor turns back into tiny droplets of water.
6. These droplets can form ______________ in the sky.

© Macmillan/McGraw-Hill

Name ______________________________ Date ____________

LESSON
Outline

What is the water cycle?

7. The ______________ shows how Earth's water evaporates to form bodies of water, and then condenses.

8. When water is warmed by the ______________ , it evaporates.

9. ______________ form when the water vapor in the air condenses.

10. Rain and ______________ then fall, and the water flows back to the oceans, rivers, and streams.

Critical Thinking

11. If there were no oceans, streams, rivers, or lakes, do you think it would still rain? Why or why not?

____

____

____

____

____

____

© Macmillan/McGraw-Hill

LESSON
Vocabulary

Name ______________________________ Date ____________

The Water Cycle

Describe what happens in each step of the water cycle.

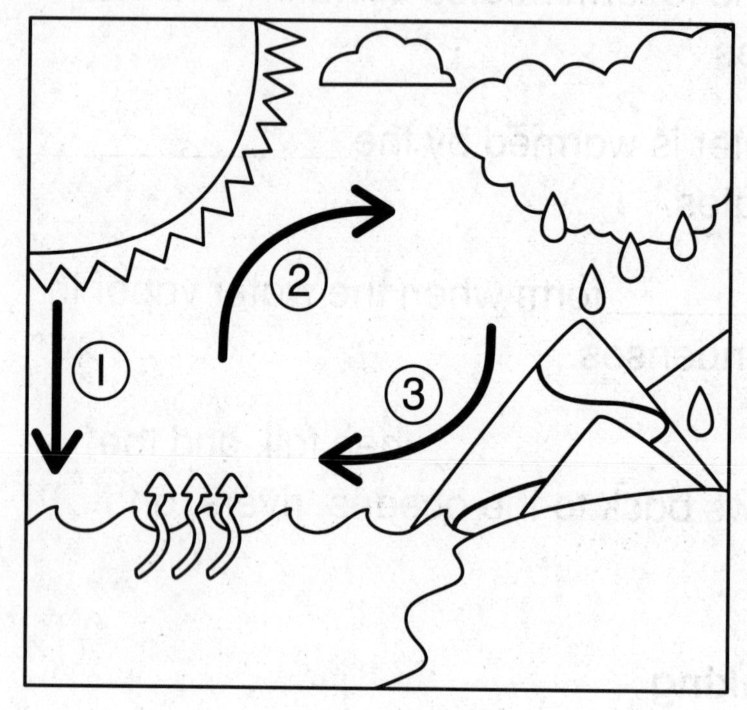

____

____

____

____

____

© Macmillan/McGraw-Hill

Name ______________________ Date ____________

The Water Cycle

Fill in the blanks. Use the words from the box.

clouds	flows	water cycle
condenses	rain	water vapor
evaporates	Sun	

How does water change? The ______________ shows how water moves from Earth to the sky, and back down again. The ______________ warms the water in oceans, rivers, and streams. The water ______________, or turns into a gas and rises. This gas is called ______________. When the air gets cooler, the water ______________, or turns back into a liquid. Tiny droplets of water form ______________ in the sky.

Precipitation like ______________ and snow can fall from the clouds. The water ______________ down the land and into the oceans, rivers, and streams. Then the cycle begins again.

© Macmillan/McGraw-Hill

LESSON **Outline**

Name ______________________________ Date ____________

Changes in Weather

Use your book to help you fill in the blanks.

What are different kinds of clouds?

1. Clouds can tell about changes in the ______________ .
2. Small, puffy clouds that can appear in long rows are called ______________ clouds.
3. Cumulus clouds are commonly seen in the ______________ .
4. Thin clouds that are very high in the sky are called ______________ clouds.
5. Cirrus clouds are made of ______________ .
6. Thick or thin clouds that are very low in the sky are called ______________ clouds.

© Macmillan/McGraw-Hill

Name ______________________ Date __________

Changes in Weather

Fill in the blanks. Use the words from the box.

cirrus	disasters	rows	tornado
cumulus	hurricane	stratus	weather

There are many different kinds of clouds. Clouds tell about changes in the ______________ . Small, white, puffy clouds are called ______________ clouds. They appear in long ______________ and mean fair weather. Thin clouds that are very high in the sky are called ______________ clouds. These clouds are made of ice. Thick or thin clouds that cover the entire sky are called ______________ clouds. These clouds mean that rain or snow is coming.

Weather can change when different types of air come together. Very strong storms can cause ______________ like floods. A ______________ is a storm with very strong winds. A ______________ is a column of spinning air. People can stay safe from many storms by staying indoors.

© Macmillan/McGraw-Hill

Name ______________________ Date __________

Predicting Storms

Read the Reading in Science pages in your book. As you read, pay attention to the most important ideas. List them in the chart below. Then summarize the article. Remember, when you summarize, you retell the most important ideas in the selection.

Idea #1

Idea #2

Summary

© Macmillan/McGraw-Hill

Name ______________________ Date ____________

Write About It

Summarize. How does Doppler radar work?

____

____

____

Write a paragraph that retells what you learned about why scientists try to predict the weather.

____

____

____

© Macmillan/McGraw-Hill

Name ______________________ Date __________

Observing Weather

Fill in the blanks. Use the words in the box.

condenses	precipitation	water cycle
evaporates	temperature	water vapor

1. The ______________ shows how water changes on Earth.

2. When water ______________ , it changes from a liquid to a gas.

3. When water ______________ , it changes from a gas to a liquid.

4. To find out hot or cold something is, we can measure its ______________ .

5. Rain, snow, sleet, and hail are all different kinds of ______________ .

6. When water is a gas, it is called ______________ .

© Macmillan/McGraw-Hill

Name ______________________ Date ____________

Solve each riddle.

1. I am thin and high in the sky. I am made of ice. What kind of cloud am I?

   ____

2. I am small, white, and puffy. I appear when the weather is fair. What kind of cloud am I?

   ____

3. I am low in the sky. I appear when rain or snow is on the way. What kind of cloud am I?

   ____

4. I am a tool that can measure the speed of the wind. What am I?

   ____

5. I am a spinning column of air. I can cause a lot of damage. What am I?

   ____

© Macmillan/McGraw-Hill

Name ______________________ Date __________

Earth and Space

Fill in the important ideas as you read the chapter.
Use the words in the box.

axis	orbit	planet	solar system
Moon	phase	rotation	

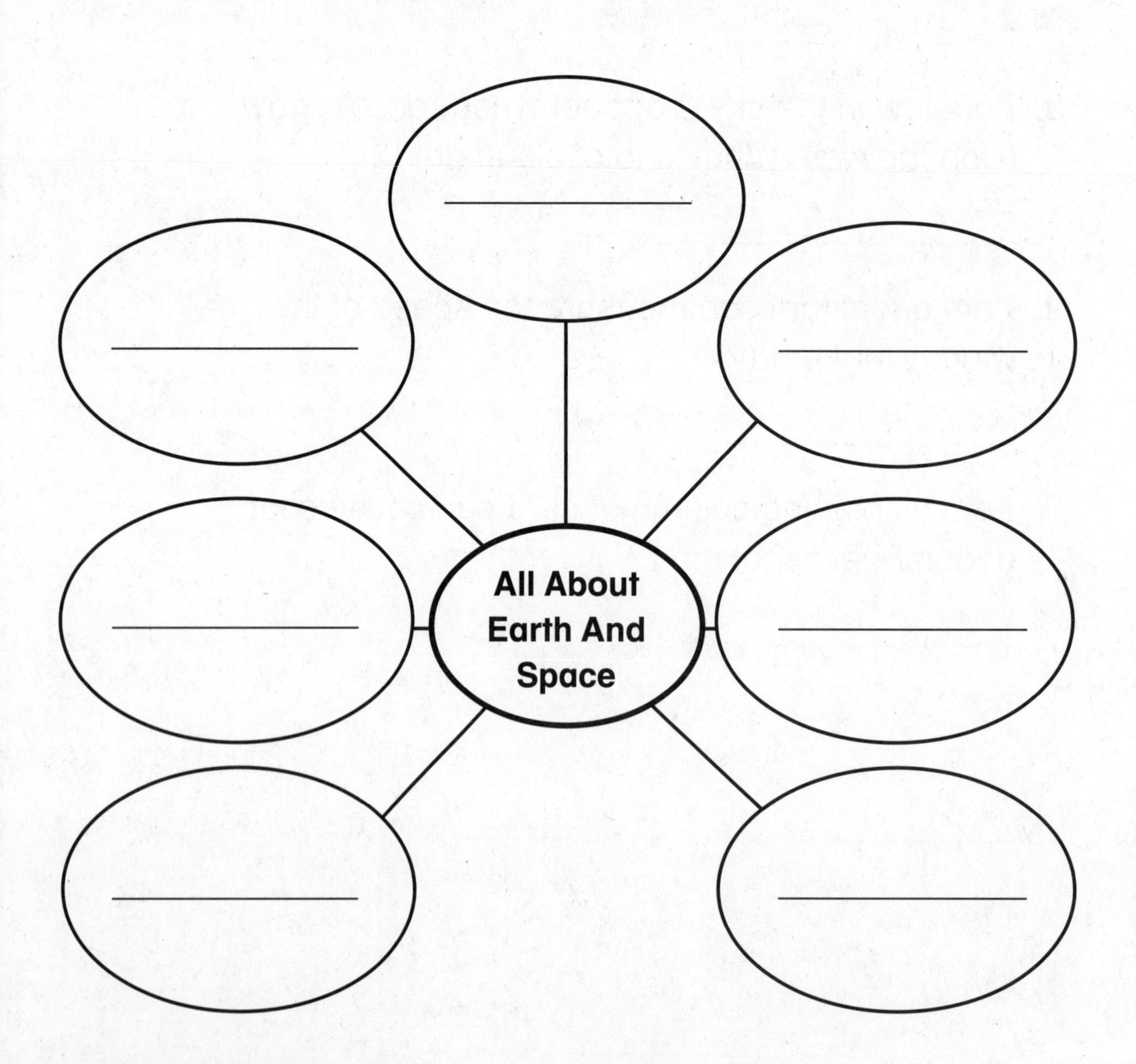

© Macmillan/McGraw-Hill

Name ______________________________ Date ___________

LESSON
Outline

Day and Night

Use your book to help you fill in the blanks.

What causes day and night?

1. Earth's ______________ is what causes day and night.

2. It is ______________ when our side of Earth faces the Sun.

3. When our side of Earth faces the Sun, it is ______________ on the other side.

4. Earth always ______________ in the same direction.

5. It takes 24 hours for Earth to make one full turn on its ______________ .

6. The axis is an imaginary line that goes through the ______________ of Earth.

© Macmillan/McGraw-Hill

Name ______________________ Date __________

Why do the Sun and Moon seem to move?

7. The ____________ seems to move across the sky during the day.

8. Shadows on the ground change as Earth __________ .

9. At night, the ____________ seems to move, too.

10. This is because ____________ is rotating.

Critical Thinking

11. What happens on the other side of Earth when it is night where you live? How do you know?

____

____

____

____

____

____

____

____

© Macmillan/McGraw-Hill

Name ______________________ Date __________

LESSON
Vocabulary

Day and Night

Fill in the blanks. Use the words from the box.

axis	day	night	rotation

1. Night and ______________ are caused by Earth's rotation.

2. Earth's ______________ never changes direction.

3. Every 24 hours, Earth rotates once on its ______________ .

4. When it is day where you live, it is ______________ on the other side of the world.

© Macmillan/McGraw-Hill

LESSON
Cloze Activity

Name ______________________ Date ____________

Day and Night

Fill in the blanks. Use the words from the box.

axis	Earth	rotation	Sun
day	night	shadows	

You cannot feel it, but you are spinning right now. In fact, ____________ is always spinning. It spins all ____________ and all night. It even spins when you are asleep! This turning is called Earth's ____________ . It is why we have day and ____________ .

Every 24 hours, Earth rotates one time on its ____________ . As it rotates, light from the ____________ lights a different part of the planet. This is why ____________ are longer during the day. When it is day on one side of the world, it is night on the other side.

© Macmillan/McGraw-Hill

Name ______________________ Date __________

LESSON
Outline

Why Seasons Happen

Use your book to help you fill in the blanks.

What are the seasons like?

1. In the fall, the ______________ is cool.

2. Some leaves ______________ colors and fall off their trees.

3. The air is much colder during the ______________ .

4. In some places, the cold rain turns to ______________ .

5. Some animals, like birds, ______________ to warmer places.

6. People wear warmer ______________ .

7. In the spring, ______________ days help new plants grow.

8. Summer is the warmest ______________ of all!

© Macmillan/McGraw-Hill

Name ______________________ Date ____________

What causes the seasons?

9. Earth takes about 365 days to ______________ the Sun.

10. Earth's orbit is its ______________ around the Sun.

11. When our part of Earth is tilted ______________ the Sun, we have spring and summer.

12. When our part of Earth is tilted ______________ from the Sun, we have fall and winter.

Critical Thinking

13. Why does the weather change during the year?

____

____

____

____

____

____

____

© Macmillan/McGraw-Hill

Name ______________________________ Date ____________

Fun with the Seasons

Write About It

Think about the seasons and the different activities you do throughout the year.

On a separate piece of paper, write a story about the activities you do in winter and in summer. Include details about how the seasons are different.

Getting Ideas

Fill in the chart with ideas about summer and winter.

Winter	Summer	Both

Planning and Organizing

Lisa wrote two sentences about winter and summer. Write Alike if the sentence shows how they are alike. Write Different if it shows how they are different.

1. ______________ Winter and summer are seasons.

2. ______________ Winter can be very cold, and summer can be very hot.

© Macmillan/McGraw-Hill

Name ____________________ Date __________

Drafting

Write a sentence to begin your paragraph. Tell how you feel about winter and summer.

____

Now write your story on a separate piece of paper. Tell what you do in winter and summer. Tell how the seasons are different.

Revising and Proofreading

Lisa wrote some sentences. She made six mistakes. Find the errors. Then correct them.

I really like winter? I like to go ice skating on the pond. I also like to go sleding. My favorite season is summer. It gets hot so I go to the beech every day with my friends. We look for shels. At night, we look at the stars and we try too find the Big Dipper. There are many activities to do in both seasons.

Now revise and proofread your writing. Ask yourself:

- ▶ Did I tell about what I like to do in winter?
- ▶ Did I tell about what I like to do in summer?
- ▶ Did I correct all mistakes?

© Macmillan/McGraw-Hill

Name ______________________ Date __________

LESSON
Outline

The Moon and Stars

Use your book to help you fill in the blanks.

Why can we see the Moon from Earth?

1. The Moon does not shine like the __________ .
2. We see the __________ of the Sun reflected off of the Moon.
3. The Moon is many __________ away from Earth.
4. The Moon is made of __________ and covered with dust.
5. The __________ helps the Moon look bright when the Sun shines on it.

Why does the Moon seem to change shape?

6. It takes the Moon about one __________ to move around Earth.
7. The Moon's __________ seems to change every few days.

© Macmillan/McGraw-Hill

Name ______________________________ Date ____________

8. The different shapes we see during the month are called ______________ of the Moon.

What are stars?

9. Stars are space objects made of hot ______________ .

10. Stars can have different ______________ and sizes.

11. Some stars make ______________ in the sky.

12. The Sun is a ______________ that gives light and heat to Earth.

Critical Thinking

13. Why can we see both the Moon and stars in the night sky?

____

____

____

____

____

____

© Macmillan/McGraw-Hill

Name ______________________________ Date ____________

LESSON
Outline

The Solar System

Use your book to help you fill in the blanks.

What goes around the Sun?

1. Earth is a ______________ .

2. Planets are huge ______________ that move around the Sun.

3. Eight planets, their moons, and the ______________ make up our solar system.

4. Like ______________ , each planet in our solor system orbits the Sun.

5. The planets that are ______________ to the Sun take less time to move around it.

What are the planets like?

6. The closest planet to the Sun is ______________ .

© Macmillan/McGraw-Hill

Name ______________________ Date ____________

7. Our planet has ______________ that we can drink and air that we can breathe.

8. Mars has a ______________, rocky surface and two moons.

9. The largest planet is called ______________ .

Critical Thinking

10. Why do you think our group of planets is called a solar system?

____

____

____

____

____

____

____

____

____

____

© Macmillan/McGraw-Hill

Name ______________________________ Date ____________

Starry, Starry Night

Read the Reading in Science pages in your book. As you read, pay attention to important ideas. How did ancient sailors find the North Star? What did they do first? What did they do last? Write your ideas in the chart below.

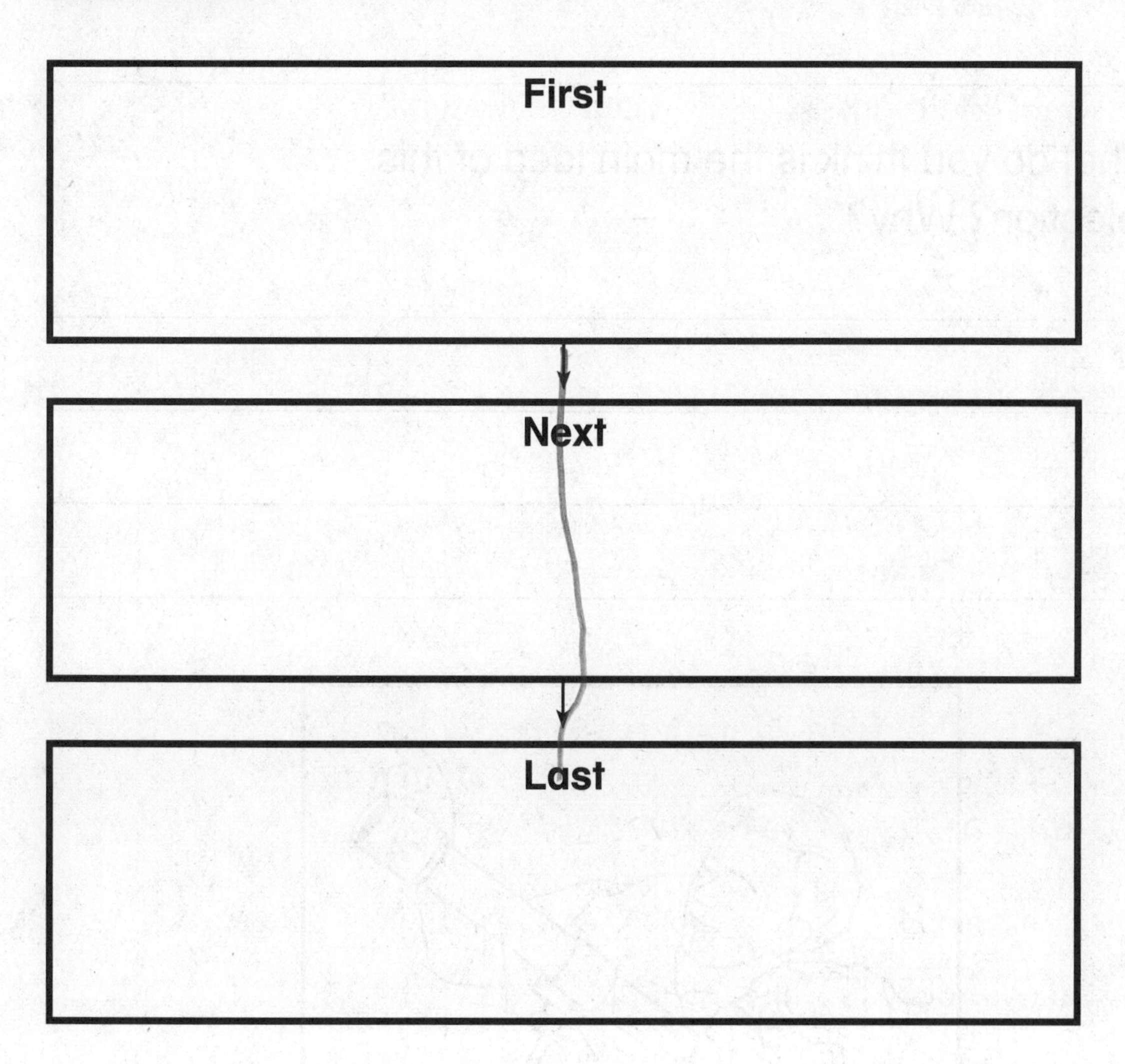

© Macmillan/McGraw-Hill

Name ______________________ Date __________

Write About It

Sequence. Long ago, sailors used star charts to find their way on the ocean. How do astronomers use star charts now?

____

____

1. What do you think is the main idea of this selection? Why?

____

____

____

____

© Macmillan/McGraw-Hill

Name ______________________ Date ____________

Popcorn Hop

by Stephanie Calmenson

Read the Unit Literature pages in your book.

Write About It

Response to Literature

1. What makes the popcorn hop?

____

____

2. How do you think popcorn got its name?

____

____

3. How do living things use heat?

____

____

____

© Macmillan/McGraw-Hill

CHAPTER
Concept Map

Name ______________________ Date ____________

Looking at Matter

Fill in the important ideas as you read the chapter. Write three facts about the properties of each kind of matter.

Matter is ______________________________________.

What Are the Properties of Matter?		
Solid	**Liquid**	**Gas**
1. ____________ ____________	**1.** ____________ ____________	**1.** ____________ ____________
2. ____________ ____________	**2.** ____________ ____________	**2.** ____________ ____________
3. ____________ ____________	**3.** ____________ ____________	**3.** ____________ ____________

© Macmillan/McGraw-Hill

Name ______________________ Date ____________

Describing Matter

Use your book to help you fill in the blanks.

What is matter?

1. Matter is anything that takes up ______________ and has mass.
2. Some matter can be ______________ by people.
3. An object's mass is the amount of ______________ it has.
4. Objects can be made of ______________ amounts of matter.
5. A ______________ is used to measure and compare mass.

How can you describe matter?

6. Matter can be described by talking about its ______________ .
7. A ______________ is how matter looks, feels, smells, tastes, or sounds.

© Macmillan/McGraw-Hill

Name ______________________ Date __________

8. Different ______________ of matter have different properties.

9. Matter can be ______________ or nonliving.

10. There are ______________ main kinds of matter: solids, liquids, and gases.

Critical Thinking

11. What are some ways that matter can be described? What do these ways tell you about matter?

____

____

____

____

____

____

____

____

____

____

© Macmillan/McGraw-Hill

Name ______________________________ Date ____________

Describing Matter

What is the secret answer? Fill in the missing words and then fill in the answer by using the circled letters.

1. Matter can be _ (_) _ _ _ or thin.

2. Anything that takes up space and has mass is called _ _ (_) _ _ _.

3. Matter can be a _ _ _ (_) _, liquid, or gas.

4. Matter can be natural or made by _ (_) _ _ _ _.

5. The amount of matter in an object is called _ (_) _ _.

6. A _ _ _ _ _ _ (_) _ describes how matter looks, feels, smells, tastes, or sounds.

Q: What did the doctor say to the scientist?

A: W _ a _ _ s th _ m _ _ t e r?

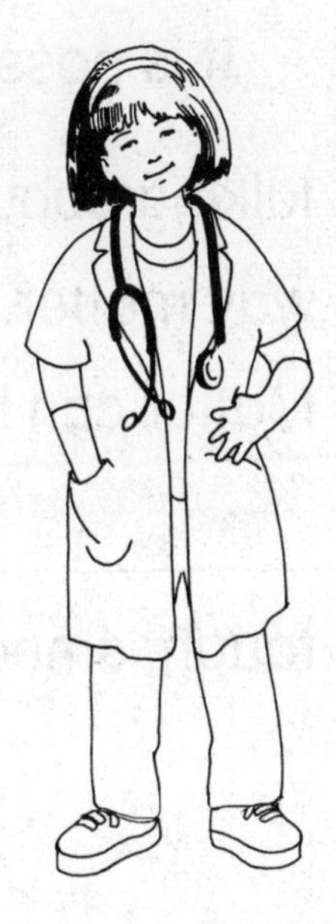

© Macmillan/McGraw-Hill

Name ______________________ Date ____________

Describing Matter

Fill in the blanks. Use the words from the box.

balance	feel	gas	matter	smaller
describe	flexible	mass	property	

Matter is everywhere. Matter can be a solid, a liquid, or a ______________. Anything that takes up space and has ______________ is matter. The amount of ______________ in an object is called mass. A ______________ can be used to measure and compare the mass of objects. Sometimes, a ______________ object has more mass than a larger object.

It is possible to ______________ matter by talking about its properties. A ______________ is a way matter looks, feels, smells, tastes, or sounds. Matter can be soft or it can be hard. Matter can be ______________ or stiff. It can also ______________ rough, smooth, or wet. Some matter is even invisible!

© Macmillan/McGraw-Hill

Name ______________________ Date ____________

LESSON
Outline

Solids

Use your book to help you fill in the blanks.

What is a solid?

1. A ______________ is one of three kinds of matter.
2. Solids have a ______________ of their own.
3. Like all matter, different solids have ______________ properties.
4. Solids can be made from ______________ like wood, plastic, and metal.
5. They can feel smooth, rough, soft, or hard when you ______________ them.

How can we measure solids?

6. Many ______________ can be used to measure solids.
7. A ______________ can be used to measure the width, length, or height of an object.

© Macmillan/McGraw-Hill

Name ______________________________ Date ____________

8. Rulers can be used to measure the lengths of objects in ____________ or inches.

9. A ____________ is used to tell how much mass something has.

10. To tell the difference between two objects, their measurements can be ____________ .

Critical Thinking

11. What will happen to a balance if you put a brick on one side and a feather on the other? Why?

____

____

____

____

____

____

____

____

© Macmillan/McGraw-Hill

Name ______________________________ Date ____________

Natural or Made by People?

Read the Reading in Science pages in your book. As you read, pay attention to important ideas. Summarize them in the chart below. Remember, when you summarize, you retell the most important ideas in the selection.

Summary

How are natural solids and humanmade solids the same and different?

____

____

____

Idea #1	Idea #2	Idea #3
________	________	________
________	________	________
________	________	________
________	________	________
________	________	________
________	________	________

© Macmillan/McGraw-Hill

Name ______________________ Date ____________

Write About It

Summarize. How is a plastic chair made? Use the chart you made to write your answer.

____

____

____

____

____

1. What are some plastic things in your classroom?

____

© Macmillan/McGraw-Hill

Name ______________________ Date ____________

LESSON **Outline**

Liquids and Gases

Use your book to help you fill in the blanks.

What is a liquid?

1. The opposite of ____________ matter is solid matter.
2. Unlike most solids, a liquid can take the shape of the ____________ it is in.
3. You can measure the ____________ of a liquid by using a measuring cup.
4. Volume is a measure of the amount of ____________ something takes up.

What is a gas?

5. A ____________ is like a liquid in many ways.
6. A gas has no ____________ of its own.
7. A bubble is liquid with ____________ inside it.

© Macmillan/McGraw-Hill

Name ______________________________ Date ____________

8. You can ______________ the volume or the mass of a gas.

9. The ______________ around us is made of many gases.

10. You can feel these gases moving on a ______________ day.

11. We need a gas called ______________ to survive.

Critical Thinking

12. What solids, liquids, and gases do you use every day?

____

____

____

____

____

____

____

© Macmillan/McGraw-Hill

Name ______________________ Date ____________

Liquids and Gases

Classify the words in the box based on their state of matter.

air	glass	juice	oxygen	water
apple	ice	milk	pencil	water vapor

Solids	Liquids	Gases
____	____	____
____	____	____
____	____	____
____	____	____

© Macmillan/McGraw-Hill

LESSON
Cloze Activity

Name ______________________ Date __________

Liquids and Gases

Fill in the blanks. Use the words from the box.

air	containers	liquid	plants	three
breathe	gas	oxygen	solid	

We use matter every day. Our clothes, shoes, breakfast, and even the ______________ we breathe are kinds of matter. There are ______________ kinds of matter. A ______________ is a kind of matter that has its own shape. A ______________ is a kind of matter that does not have a shape of its own. A ______________ is another kind of matter that does not have its own shape. Gases and liquids take the shapes of the ______________ they are in.

The air we ______________ is made of many gases. One of these gases in the air is called ______________. Animals and ______________ need oxygen to live. We cannot see gases but they are all around us.

© Macmillan/McGraw-Hill

Name ______________________________ Date ____________

Fun with Water

Write About It

This girl is having fun in the water! Think of times that you have had fun in water. Draw and write about what you did.

Getting Ideas

Look at the starfish. Write Water in the center. In the arms, write things you do to have fun in the water.

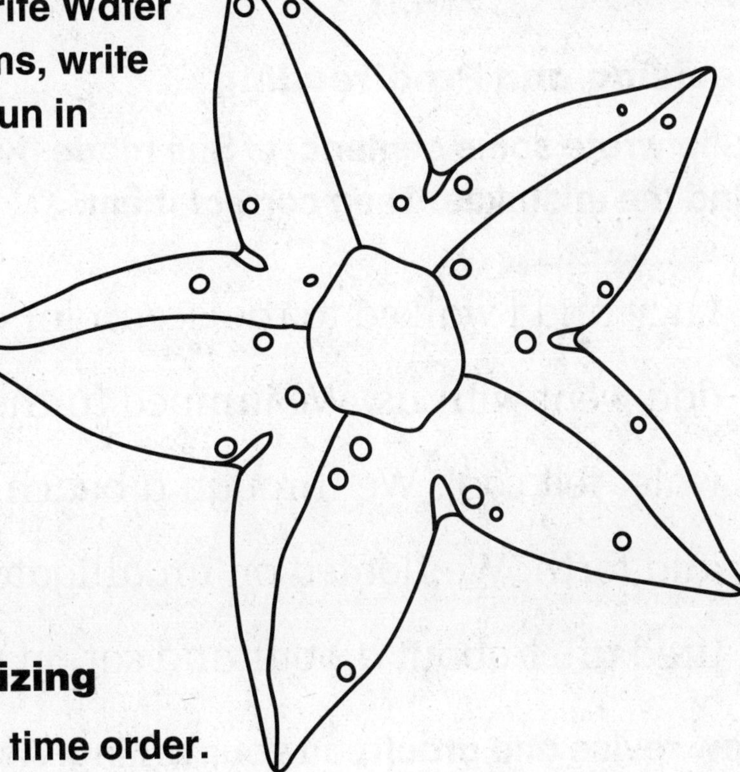

Planning and Organizing

Put these sentences in time order.

__________ I jumped into the water.

__________ I put on my bathing suit and packed some toys.

__________ My mother and I walked to the beach.

© Macmillan/McGraw-Hill

Name ______________________ Date __________

Drafting

Write a sentence to begin your story. Use I to write about yourself.

____

____

Now write your story on a separate piece of paper. Tell about fun that you have had in the water. Tell how the water made you feel.

Revising and Proofreading

Julia wrote some sentences. She made five mistakes. Find the mistakes. Then correct them.

Lucy and i walked to the ocean for a swim. His dad went with us. We jumped in the weaves. The water felt cool. We through a beach ball back and forth. We floated on an alligator raft. We got tired after about a hour and sat on our towels.

Now revise and proofread your writing. Ask yourself:

- Did I write about what I did in the water?
- Did I tell how I felt?
- Did I correct all mistakes?

© Macmillan/McGraw-Hill

Name ______________________________ Date ____________

Looking at Matter

Fill in the blanks. Use the words in the box.

balance	matter	solid
mass	property	volume

1. Anything that takes up space and has mass is ______________ .

2. The amount of matter in an object is called ______________ .

3. A ______________ can be used to measure and compare mass.

4. The amount of space something takes up is called ______________ .

5. A ______________ has a shape of its own.

6. A ______________ is how matter looks, feels, smells, sounds, or tastes.

© Macmillan/McGraw-Hill

Name ______________________________ Date ____________

Write whether each fact describes a solid, a liquid, or a gas.

1. This kind of matter has a shape of its own.

   ____

2. It can not be seen, but it is everywhere.

   ____

3. Water is an example of this kind of matter.

   ____

4. Oxygen is an example of this kind of matter.

   ____

5. This can be made of plastic, metal, or wood.

   ____

6. This kind of matter can be measured by using a measuring cup.

   ____

© Macmillan/McGraw-Hill

Name ______________________________ Date ____________

CHAPTER
Concept Map

Changes in Matter

Using what you have learned from the chapter, fill in the blanks to tell how matter can change.

Physical Change

____

____

____

____

Chemical Change

____

____

____

____

Change and Matter

Changes of State

____

____

____

____

Mixtures

____

____

____

____

© Macmillan/McGraw-Hill

LESSON **Outline**

Name ______________________ Date ____________

Matter Changes

Use your book to help you fill in the blanks.

What are physical changes?

1. Physical changes cause a ______________ in matter.

2. A physical change takes place when the size or shape of ______________ changes.

3. The ______________ of matter stays the same if its shape is changed.

4. When a piece of paper is folded or torn, a ______________ change is taking place.

5. A change in ______________ can be a physical change, too.

6. When something gets ______________ or dries, it may look and feel different, but it is only a physical change.

© Macmillan/McGraw-Hill

Name ______________________ Date __________

What are chemical changes?

7. During a ______________ change, one kind of matter changes into a different kind of matter.

8. When ______________ goes through a chemical change, it may not be possible to change it back.

9. When wood is ______________ in a fireplace, a chemical change is taking place.

10. Observing ______________ and feeling ______________ and cold are clues that a chemical change may be occuring.

Critical Thinking

11. Think about a piece of bread. How can you make a physical change to the bread? How can you make a chemical change?

____

____

____

____

© Macmillan/McGraw-Hill

Name ______________________ Date ____________

Matter Changes

Identify each description as a physical change or a chemical change.

1. An iron screw rusts in the rain.

____

2. A piece of paper is folded.

____

3. A rock breaks down into soil.

____

4. Water freezes and turns into ice.

____

5. A peach turns brown.

____

6. A ball gets wet.

____

7. A slice of cheese melts.

____

8. An egg is fried.

____

© Macmillan/McGraw-Hill

Name ______________________________ Date ____________

LESSON
Cloze Activity

Matter Changes

Fill in the blanks. Use the words from the box.

burns	mass	rusts
chemical change	matter	temperature
fold	physical change	

Matter changes every day. A ______________ takes place when the size or shape of matter changes but not the type of matter. When you ______________ paper, you are making a physical change. When only the shape of an object changes, its ______________ stays the same. When the ______________ of water changes, it can freeze or boil. These are physical changes, too.

You can also make a ______________ to matter. A chemical change happens when ______________ changes into a different kind of matter. When matter ______________ , it can not change back to its original form. When iron ______________ , it changes color and feels different. These are chemical changes at work.

© Macmillan/McGraw-Hill

Name ______________________ Date __________

Changes of State

Use your book to help you fill in the blanks.

How can heating change matter?

1. Heat can change ____________ in different ways.
2. When a solid gets enough ____________ , it melts.
3. When something melts, it changes from a ____________ to a liquid.
4. When heat is added to ice, it turns into ____________ water.
5. Different solids can ____________ at different temperatures.
6. Some liquids ____________ when they get enough heat.
7. When a liquid boils, it ____________ , or changes into a gas.
8. This gas is called ____________ .

© Macmillan/McGraw-Hill

Name ______________________ Date ____________

LESSON
Outline

How can cooling change matter?

9. When you ____________ matter, you take heat away from it.

10. A gas can ____________ when it is cooled.

11. When a ____________ condenses, it changes into a liquid.

12. When ____________ lose enough heat, they freeze.

13. When matter ____________, it changes from a liquid to a solid.

Critical Thinking

14. Explain how you can make an ice cube change from a solid to gas.

____

____

____

____

____

____

© Macmillan/McGraw-Hill

LESSON
Vocabulary

Name ________________________ Date __________

Changes of State

Solve the riddles and fill in the puzzle.

Down

1. I keep my shape when I'm cool. If it gets too warm, I melt. ______________
2. You can add me or take me away to change matter. ______________
4. This happens when liquids get very cold. ______________
6. When I start out very hot and then become cool, I turn into liquid. ______________

Across

3. This is what gas does when 6 Down happens. ______________
5. This is how solids turn into liquids. ______________
7. This is how matter goes into the air when it boils. ______________

1.
2.
3.
4.
5.
6.
7.

© Macmillan/McGraw-Hill

Name ______________________________ Date ____________

LESSON
Cloze Activity

Changes of State

Fill in the blanks. Use the words from the box.

condense	heat	solid
evaporate	liquid	temperatures
freeze	melt	water vapor

There are three main states, or forms, of matter. The three main states are ______________, liquid, and gas. Some solids ______________ when they get enough heat. When something melts, it changes from a solid to a ______________. That is what happens when an ice cube melts. Different solids melt at different ______________. When water boils, it will ______________, or turn into a gas. This gas is called ______________.

When ______________ is taken away from matter, it can change. Gases ______________ when they are cooled. When you ______________ water, it turns into a solid. Different liquids freeze at different temperatures.

© Macmillan/McGraw-Hill

Name ______________________ Date __________

Colorful Creations

Read the Reading in Science pages in your book. Write inferences based on the statements in the "What I Know" column. Write your inferences on the chart.

What I Know	What I Infer
Most crayons are made of wax. Colored wax is melted into a liquid.	
The crayon mold is cooled with cold water.	
A machine packs the crayons into boxes.	

© Macmillan/McGraw-Hill

Name ______________________ Date ____________

Write About It

Predict. What do you think would happen if the mixture of wax was poured into a mold shaped like a square? Explain your answer.

____

____

____

____

____

What two states of matter are used to make crayons?

____

How do you think different colored crayons are made?

____

____

© Macmillan/McGraw-Hill

LESSON
Outline

Name ______________________________ Date ____________

Mixtures

Use your book to help you fill in the blanks.

What are mixtures?

1. When two or more things are put together, the result is called a ______________ .
2. Mixtures can have different ______________ of solids, liquids, and gases.
3. Some mixtures can be picked ______________ .

Which mixtures stay mixed?

4. A mixture that is difficult to take apart is called a ______________ .
5. When salt is added to water, the salt ______________ and mixes with the water.
6. Sand and water ______________ make a solution.

© Macmillan/McGraw-Hill

Name ______________________________ Date ____________

LESSON
Outline

How can you take mixtures apart?

7. Some mixtures are ______________ to take apart. Other mixtures are more difficult.

8. A ______________ can be used to separate sand from water.

9. A ______________ can be used to separate iron from sand.

10. To take out salt from salt water, a process called ______________ is used.

Critical Thinking

11. Suppose you had a mixture of water and pebbles. How could you take apart the mixture?

____

____

____

____

____

____

____

____

© Macmillan/McGraw-Hill

Name ______________________ Date __________

Mixtures

Write whether you would need to use a magnet, a filter, evaporation, or your hands in order to take apart each mixture listed below. Some mixtures can be taken apart in more than one way.

1. salt water

____

2. water and sand

____

3. iron nails and sand

____

4. raisins and cornflakes

____

5. iron screws and plastic beads

____

6. pennies and nickels

____

7. blue paper and white paper

____

8. water and seashells

____

© Macmillan/McGraw-Hill

Name ______________________________ Date ____________

LESSON
Cloze Activity

Mixtures

Fill in the blanks. Use the words from the box.

dissolves	filter	magnet	separate
evaporation	liquids	mixture	solution

Have you ever made a collage? When you glue pieces of paper together, you make a ______________. A mixture can be any combination of solids, ______________, and/or gases. Some mixtures can be ______________ by their parts.

When salt and water are mixed together, a ______________ is made. The salt can not be seen because it ______________ in the water. The mixture can be taken apart by using ______________. The water will evaporate and the salt will be left behind. To separate water and sand, a ______________ can be used. To separate iron and sand, a ______________ can be used. You can separate some mixtures by using your hands.

© Macmillan/McGraw-Hill

Name ______________________________ Date ____________

Writing a Recipe

Write About It

You can write a recipe. Explain how you would use some of this fruit to make a fruit salad. Explain why it is a mixture.

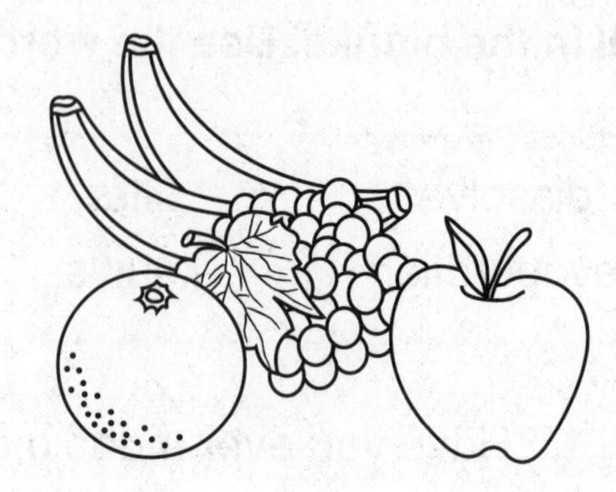

Getting Ideas

Look at the illustration. What kinds of fruit do you see? Think about how you would make a fruit salad.

What kinds of fruit would you want to put in a fruit salad? List them below.

____

Planning and Organizing

Put the steps in the correct order.

__________ Mix the fruit together.

__________ Wash the fruit and put it on the cutting board.

__________ Get a bowl and a cutting board.

__________ Cut up each fruit. Put the fruit in the bowl.

© Macmillan/McGraw-Hill

Name ______________________________ Date ____________

Drafting

Write a sentence to begin your recipe. Tell what the recipe is for.

____

____

Now write the recipe on a separate piece of paper. Put the steps in order. At the end, tell why it is a mixture.

Revising and Proofreading

Use the words in the box to fill in the blanks.

Finally	First	Next	Second	Then

______________, I put a big bowl on the counter.

______________, I got a spoon. ______________,
I put cut-up apples and bananas in the bowl. ______________,
I added grapes, blueberries, and strawberries.

______________, I mixed everything together.

Now revise and proofread your writing. Ask yourself:

- ▶ Did I write the steps in order?
- ▶ Did I explain why it is a mixture?
- ▶ Did I correct all mistakes?

© Macmillan/McGraw-Hill

Name ______________________ Date __________

Changes in Matter

Fill in the blanks. Use the words in the box.

chemical change	evaporation	melts
condenses	freezes	solution

1. When matter ______________, it changes from a solid to a liquid.
2. A process called ______________ can be used to separate salt from water.
3. A ______________ is a mixture that is difficult to separate.
4. When matter ______________, it changes from a gas to a liquid.
5. When water ______________, it changes from a liquid to a solid.
6. When a slice of bread is toasted, a ______________ occurs.

© Macmillan/McGraw-Hill

Name ______________________ Date __________

Draw a line from each picture to the sentence that describes it.

1\. 

2\. 

3\. 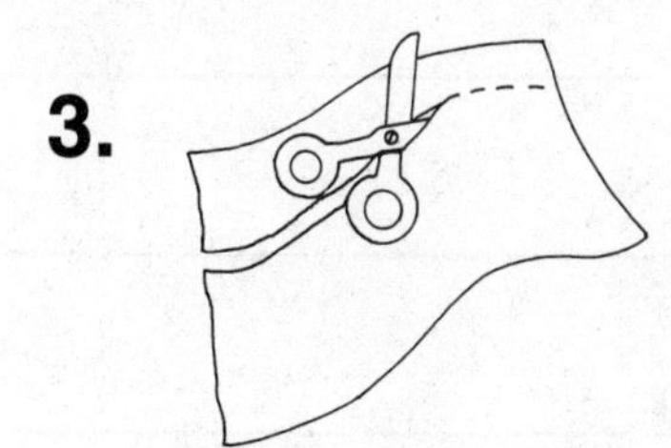

4\. 

5\. 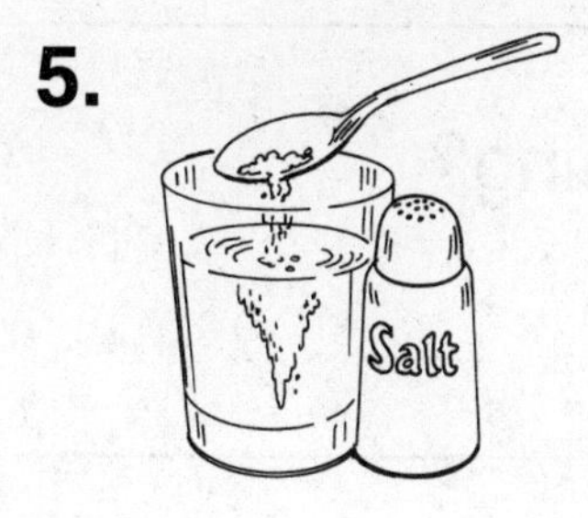

a. Salt dissolves in water to make a solution.

b. When a physical change takes place, matter changes shape.

c. Evaporation is when matter changes from a liquid to a gas.

d. After a chemical change takes place, matter may look and smell different than before.

e. When matter melts, it changes from a solid to a liquid.

© Macmillan/McGraw-Hill

Name ______________________ Date __________

Echolocation

Read the Unit Literature pages in your book.

Write About It

Response to Literature

1. Why do you think that bats use echolocation? Use the article to tell how you know.

____

____

____

____

2. What other animals do you think use echolocation?

____

____

3. Have you ever used sound to find something? Write about it.

____

____

____

© Macmillan/McGraw-Hill

Name ______________________ Date ____________

How Things Move

Fill in the important ideas as you read the chapter.

How do things move?		
Where things move	**What makes things move**	**Ways things move**

© Macmillan/McGraw-Hill

LESSON
Outline

Name ______________________________ Date ____________

Position and Motion

Use your book to help you fill in the blanks.

What are position and motion?

1. You can use ______________ words to describe an object's location.

2. Position is the ______________ where something is.

3. Above, ______________, left, and right are all position words.

4. When an object ______________, it changes position.

5. When an object is moving, it is in ______________.

6. You can ______________ the position and motion of objects.

© Macmillan/McGraw-Hill

Name ______________________________ Date ____________

LESSON
Cloze Activity

Position and Motion

Fill in the blanks. Use the words from the box.

compare	motion	space
left	object	stopwatch
measure	position	

How do you know where something is? We use words like above, below, ______________, and right to describe where things are. When you describe an object's ______________, you tell where it is. To tell the position of an object, you can ______________ it to another object.

Objects do not always stay in the same place. When an ______________ moves, its position changes. This is called ______________. Speed is a ______________ of how quickly an object moves from one position to another. The ______________ between the two positions is called distance. You can use a ______________ to measure speed. You can use a tape measure to measure distance.

© Macmillan/McGraw-Hill

LESSON
Outline

Name ______________________ Date __________

Forces

Use your book to help you fill in the blanks.

What makes things move?

1. It takes a ______________ or a pull to make something move.

2. A push or pull is a ______________ .

3. To push something, you move it ______________ you.

4. To pull something, you move it ______________ you.

What are some forces?

5. When you throw a ball in the air, ______________ pulls it back to Earth.

6. Gravity is a force that ______________ things to Earth.

7. One ______________ of gravity is weight.

8. ______________ is how much force it takes to pull something to Earth.

9. When this happens, a force called ______________ slows down the objects.

© Macmillan/McGraw-Hill

Name ______________________________ Date ____________

LESSON
Outline

How can forces change motion?

10. Forces can make things ____________ up, slow down, or change direction.

11. Sometimes, objects ____________ together when they move.

Critical Thinking

12. Do you think gravity is important? Why or why not?

____

____

____

____

____

____

____

____

© Macmillan/McGraw-Hill

Name ____________________ Date __________

Forces

Answer each riddle. Then find each word in the word search.

1. I am a force that slows down moving things.

   What am I? ______________

2. I am a force that pulls things to Earth.

   What am I? ______________

3. To put an object in motion, you must use me.

   What am I? ______________

4. I am the amount of force that pulls an object to Earth. What am I? ______________

5. To move an object closer to you, you must use me. What am I? ______________

f	r	i	c	t	i	o	n	d	w
o	l	m	s	h	i	e	h	g	e
r	g	r	a	v	i	t	y	c	i
c	a	t	v	m	p	s	t	u	g
e	m	n	x	y	r	l	m	e	h
p	u	l	l	n	z	c	b	o	t

© Macmillan/McGraw-Hill

Name ______________________ Date ___________

LESSON
Cloze Activity

Forces

Fill in the blanks. Use the words from the box.

amount	down	pull
away	force	push
direction	gravity	

How do you move things? Think about the last time you threw a ball. You used a ____________ to move the ball. A force is a ____________ or pull that makes objects move. When you ____________ an object, you move it closer to you. When you push an object, it moves ____________ from you.

You can use forces to speed up or slow ____________ an object. Friction is a force that slows some things down. Forces can even change the ____________ of an object's motion. The force that pulls objects to Earth is called ____________ . The ____________ of force that gravity pulls down on an object is called weight. People use forces every day.

© Macmillan/McGraw-Hill

Name ______________________________ Date ____________

Meet Hector Arce

Read the Reading in Science pages in your book. As you read, keep track of what happens and why. Record the causes and effects you read about in the chart below. Remember, a cause is why something happens. An effect is the thing that happens. Sometimes, one cause can have many effects.

Cause	Effect
Gravity	
	It pulls together huge clouds of gas and dust to form stars.
Gravity	

© Macmillan/McGraw-Hill

Name ________________________________ Date ____________

Use the words in the box to retell what you learned about the effects of gravity.

dust	gas	hot
force	gravity	stars

The ______________ that pulls objects toward Earth is called ______________. It keeps all living things and objects on Earth as the planet spins. Gravity also pulls on other planets and on moons. It can even cause ______________ to form. Gravity pulls together clouds of ______________ and ______________ to make stars. Inside these stars, gravity makes them so ______________ that they glow in the night sky.

Write About It

Cause and Effect. What causes stars to form?

____

____

____

© Macmillan/McGraw-Hill

LESSON
Outline

Name ______________________________ Date ____________

Using Simple Machines

Use your book to help you fill in the blanks.

What are levers and ramps?

1. A ______________ is a tool that can change the strength of a force.
2. A ______________ is a simple machine with a bar that moves on a stationary fulcrum.
3. This machine can change how much force is needed for a ______________ so you can move heavy things.
4. A seesaw and a ______________ are kinds of levers.
5. A ______________ is another kind of simple machine that can help you move things.
6. A ramp has a ______________, slanted surface.

© Macmillan/McGraw-Hill

Name ______________________________ Date ____________

LESSON
Outline

What are other simple machines?

7. People use simple machines like axles and

_______________ every day.

8. An _______________ is a bar that is connected to the center of a wheel.

9. A simple machine made of a rope that moves

around a _______________ is called a pulley.

10. Pulleys make it easier to _______________ things.

Critical Thinking

11. Where have you seen ramps? Why are these and other simple machines useful?

____

____

____

____

____

____

© Macmillan/McGraw-Hill

LESSON
Vocabulary

Name ______________________ Date ____________

Using Simple Machines

Identify the simple machine in each picture.

1.

____

2.

____

3.

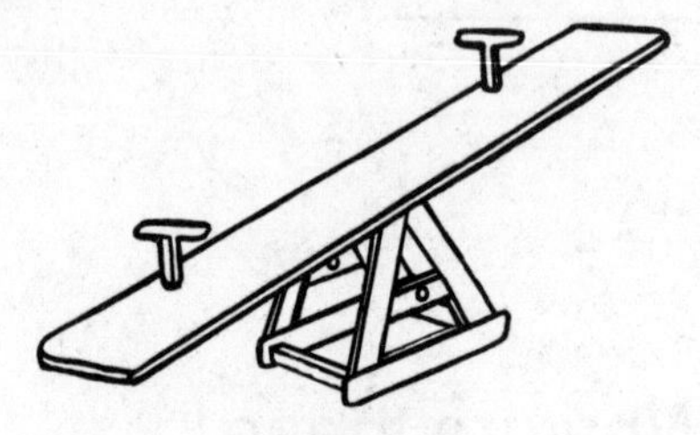

____

4.

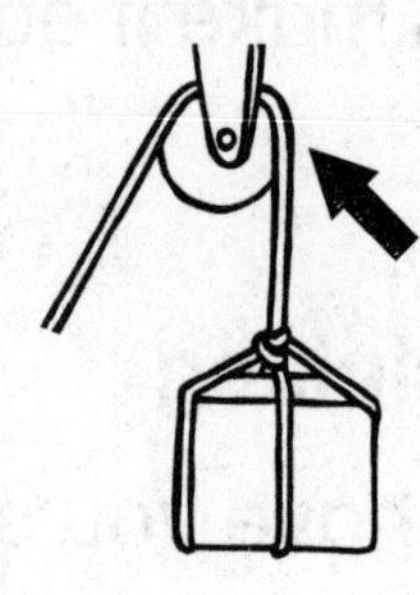

____

5.

____

© Macmillan/McGraw-Hill

Name ______________________________ Date ____________

LESSON
Cloze Activity

Using Simple Machines

Fill in the blanks. Use the words from the box.

axle	lever	simple machine
force	pulley	
fulcrum	ramp	

Tools help people change the ______________ used on an object. Sometimes, objects are too heavy to lift or move on our own. A ______________ is a tool that can change the size or direction of a force. A ______________ is a bar that moves on a point that stays still. This point is called a ______________ . People use this tool to lift heavy things. A ______________ is used to move things from one place to another. We can push objects on its slanted surface.

Cars and bikes have wheels that help them move. An ______________ is a bar connected to the center of the wheel. A ______________ has a rope that moves around a wheel. This tool helps people change the direction of an object.

© Macmillan/McGraw-Hill

Name ______________________ Date ____________

Slip and Slide

Write About It

Explain why penguins can slide on the ice.
Think about what you learned about forces.
Make sure to explain why ice is slippery.

Getting Ideas

Brainstorm a list of facts about penguins, and write them in the chart below.

Penguin Fact Sheet

Planning and Organizing

Zina wrote four sentences. Write Yes if the sentence is a penguin fact. Write No if it is not.

1. ________ Penguins are birds that have webbed feet.
2. ________ Penguins have black and white feathers.
3. ________ Penguins can fly.
4. ________ Penguins have short legs.

© Macmillan/McGraw-Hill

Name ______________________ Date __________

Drafting

Write your own topic sentence to begin your paragraph. Tell your main idea about penguins.

____

____

Now write about penguins on a separate piece of paper. Start with your main idea. Explain how they slide on the ice. Tell which body parts help them move.

Revising and Proofreading

Zina wrote a paragraph. She made five mistakes. Find the mistakes. Then correct them.

Penguins slide on their bellys. They use their feet and flipers. Their feet push them forward. There flippers balance them. When they glide, the ice under them melts. This makes the ice slipperie. They can glide a few miles an our. Gliding takes less energy than walking.

Now revise and proofread your writing. Ask yourself:

- Did I follow all instructions?
- Did I correct all mistakes?

© Macmillan/McGraw-Hill

Name ______________________ Date ____________

Exploring Magnets

Use your book to help you fill in the blanks.

What do magnets do?

1. Magnets use ______________ to attract some objects.
2. Magnets can pull objects without ______________ them.
3. A ______________ can attract objects made of iron, nickel, or steel.
4. Strong magnets can ______________ objects that are far away.
5. Magnets can pull objects that contain ______________ or steel.
6. Magnets can not pull objects made of ______________ or plastic.

© Macmillan/McGraw-Hill

Name ______________________________ Date ____________

LESSON
Cloze Activity

Exploring Magnets

Fill in the blanks. Use the words from the box.

attract	magnet	north
iron	nickel	south

It is possible to move objects without even touching them. A ______________ can make some things move. It uses force to ______________ , or pull, some objects. It can pull objects that contain ______________ , like paper clips and screws. It can also pull objects that contain ______________ . A magnet can not attract things made out of wood or plastic.

Every magnet has two poles. If the ______________ pole of one magnet is put next to the south pole of another magnet, the two magnets will attract. If the ______________ pole of one magnet is put next to the south pole of another, the two magnets will repel. Magnets are powerful!

© Macmillan/McGraw-Hill

Name ______________________ Date ____________

How Things Move

Fill in the blanks. Use the words in the box.

friction	lever	position
gravity	poles	simple machine

1. A ______________ is a tool that can change the size or direction of a force.
2. A force that slows down moving things is called ______________ .
3. Every magnet has two ______________ .
4. A ______________ is a simple machine that helps people lift heavy things.
5. You can tell the ______________ of an object by comparing it to another object.
6. The force that pulls things toward the ground is called ______________ .

© Macmillan/McGraw-Hill

Name ______________________ Date __________

Complete the sentences. Then fill in the puzzle.

Down

1. When you ______________ something, you move it away from you.

3. A ______________ is a simple machine with a straight surface that is slanted.

5. The amount of force that pulls an object down toward Earth is called its ______________.

Across

2. An ______________ is a bar that is connected to the center of a wheel.

4. The point on a lever that stays still is called the ______________.

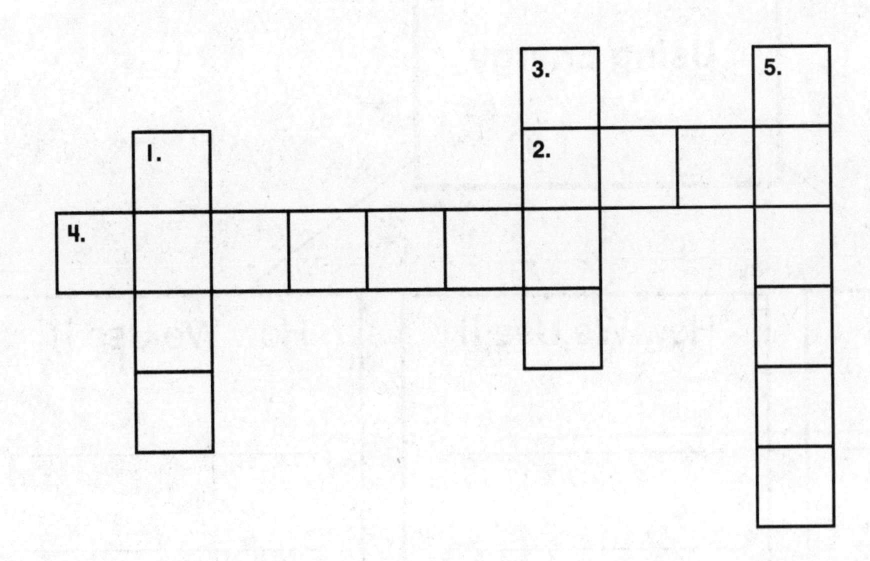

© Macmillan/McGraw-Hill

Name ______________________ Date __________

Using Energy

Fill in the important ideas as you read the chapter. Use the words in the box to fill in the first row. Use your own ideas to fill in the second row.

heat	light	sound

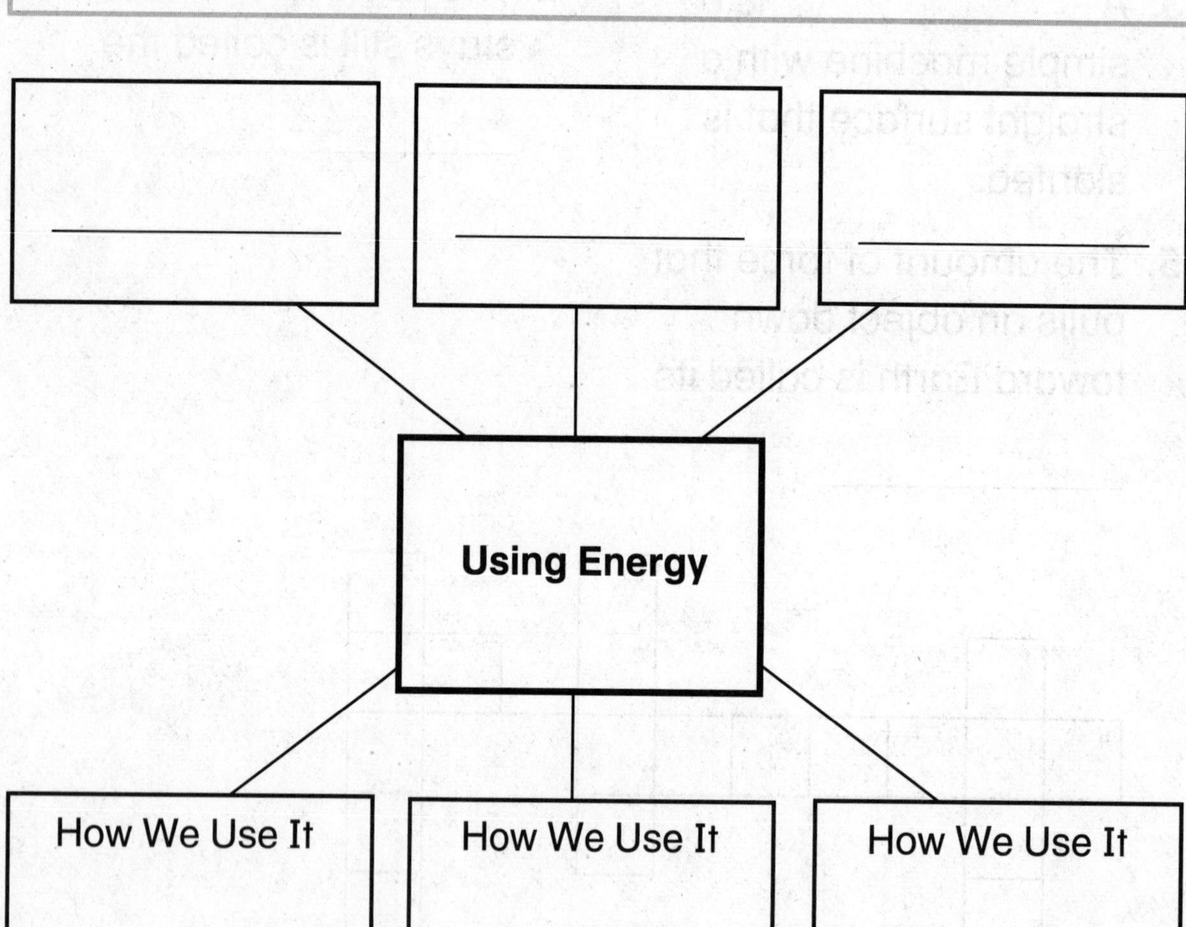

© Macmillan/McGraw-Hill

Name ______________________ Date ____________

LESSON
Outline

Heat

Use your book to help you fill in the blanks.

What is heat?

1. Energy makes ______________ move or change.
2. Heat is energy that can change the ______________ of matter.
3. Heat can ______________ solids and turn liquids into gases.
4. The ______________ gives Earth most of its heat.
5. We can also get heat ______________ from other things.
6. Something that gives off heat energy when it is burned is ______________ .
7. Heat energy can also come from ______________ .

© Macmillan/McGraw-Hill

LESSON
Outline

Name ______________________ Date ____________

What is temperature?

8. We can tell how hot or cold something is by measuring its ______________ .

9. Thermometers have a special ______________ inside of them.

10. When the temperature is ______________ , the liquid goes up.

11. When the temperature is cool, the liquid goes ______________ .

Critical Thinking

12. What are three sources of heat energy? How do we measure this energy?

____

____

____

____

____

____

© Macmillan/McGraw-Hill

Name ______________________ Date ____________

Heat

Read each sentence. Write TRUE if the sentence is true. Write NOT TRUE if the sentence is false.

1. ______________ Heat energy can change the states of matter of some objects.

2. ______________ Heat can turn a gas into a solid.

3. ______________ Most heat energy comes from the Moon.

4. ______________ Gas, oil, wood, and coal are all types of fuel.

5. ______________ Temperature is a measure of how hot or cold something is.

6. ______________ Thermometers measure how fast someone is running.

© Macmillan/McGraw-Hill

Name ______________________ Date __________

Heat

Fill in the blanks. Use the words from the box.

coal	fuel	matter
friction	heat energy	temperature

There are many elements of energy. Energy makes ______________ move or change. The Sun gives ______________ to Earth. Heat energy keeps us warm.

Not all heat energy comes from the Sun. Gas, oil, wood, and ______________ give off heat energy. Things that give off heat when burned are called ______________. You can make heat energy, too! When you rub your hands together quickly, the ______________ makes heat energy.

A measure of hot and cold is called ______________. A thermometer is a tool that people use to measure temperature.

© Macmillan/McGraw-Hill

Name ______________________________ Date ____________

LESSON **Outline**

Sound

Use your book to help you fill in the blanks.

What makes sound?

1. Another kind of energy we use every day is

   ______________.

2. When objects ______________, they give off sound energy.

3. A vibrating object moves ______________ and forth quickly.

4. When your ______________ vibrates, you hear sound.

5. Your ______________ helps you figure out what you are hearing.

How are sounds different?

6. Some sounds are ______________ and some sounds are loud.

7. Soft sounds have less energy than

   ______________ sounds.

© Macmillan/McGraw-Hill

Name ______________________ Date ____________

8. Some sounds have a higher ____________ than other sounds.

9. Pitch is how high or ____________ a sound is.

What do sounds move through?

10. Sound can ____________ through air.

11. Sound energy can even move through ____________ and many liquids!

Critical Thinking

12. How do we hear sound? How are sounds different?

____

____

____

____

____

____

____

____

© Macmillan/McGraw-Hill

Name ____________________ Date ____________

Sound

Describe what each picture shows about sound.

1.

____

____

2.

____

____

3.

____

____

© Macmillan/McGraw-Hill

LESSON
Cloze Activity

Name ______________________________ Date ____________

Sound

Fill in the blanks. Use the words from the box.

eardrum	liquids	sound
energy	pitch	vibrate

Did you know that we can hear a kind of energy? The kind of energy that we can hear is ______________. Sound energy is made when objects ______________. Sound can travel through air. Sound can also travel through solids and ______________. The closer you are to a sound, the louder it will be.

How do we hear these sounds? The part of our body we use to hear sounds is the ______________. It sends messages to our brain about what sound we heard. Not all sounds are the same. A whisper has less ______________ than a shout. The ______________ is how high or low a sound is. Imagine a guitar's strings. The tighter the strings are, the higher the pitch is. There are many different sounds.

© Macmillan/McGraw-Hill

Name ______________________________ Date ____________

Sound Off!

Write About It

Describe the pitch and volume of a sound you hear every day. How do we use sounds? Why are sounds important?

Getting Ideas

Choose a sound you hear every day. Write it in the center ovals. In the outer ovals, write words that describe that sound.

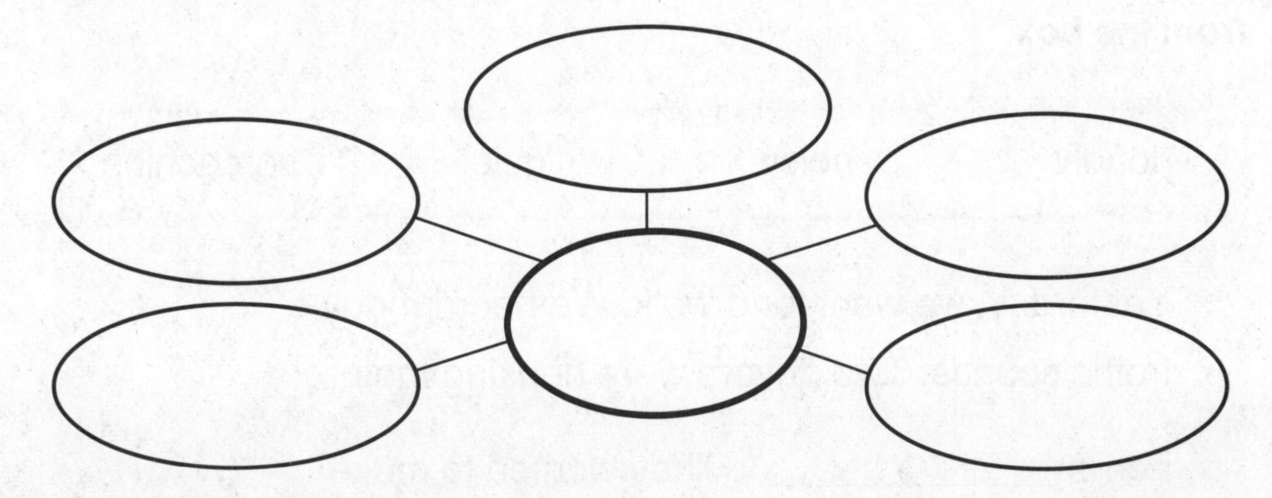

Planning and Organizing

Circle the descriptive words in these sentences.

1. The brown sparrow sang loudly.
2. The little sparrow sang a pretty song.

© Macmillan/McGraw-Hill

 Name ______________________ Date __________

Drafting

Write a sentence to begin your paragraph that tells an important idea about a sound you hear every day.

____

Write about the sound on a separate piece of paper. Remember to use descriptive words.

Revising and Proofreading

Pedro wrote a paragraph. He did not use any describing words. Fill in the blank spaces with words from the box.

loudly	noisy	quiet	screeching

Yesterday, we went for a walk. We heard many traffic sounds. Two drivers were honking their horns ______________ . They wanted to make sure a boy on a bike saw them. A car stopped at a red light. It made a ______________ sound. Then two fire engines went zooming past us. The traffic sounds were so ______________ . There was not one ______________ place in the city.

© Macmillan/McGraw-Hill

Name ______________________________ Date ____________

LESSON
Outline

Light

Use your book to help you fill in the blanks.

What is light?

1. Did you know that ______________ energy helps you see things?

2. Some light comes from ______________ and flashlights.

3. Most light on Earth comes from the ______________ .

4. Light ______________ off of objects and goes into our eyes to help us see.

5. The dark area made when something is blocking light is called a ______________ .

6. Some ______________ objects can block light and make shadows.

How do we see color?

7. White light is really a mix of different ______________ of light.

© Macmillan/McGraw-Hill

Name ______________________________ Date ____________

8. When light ____________, we can see the colors of the rainbow.

9. A ____________ is a tool that helps to bend light.

10. A ____________ is a tool that blocks some colors of light.

Critical Thinking

11. Why is light important? How many kinds of energy does the Sun give to Earth?

____

____

____

____

____

____

____

____

____

____

© Macmillan/McGraw-Hill

Name ______________________ Date __________

LESSON
Vocabulary

Light

Fill in the blanks. Use the words from the box.

colors	eyes	prism	reflects
energy	light	rainbow	

1. Light is a mix of ______________ .

2. My ______________ are important tools that let me see the world around me.

3. Heat, sound, and light are all kinds of ______________ .

4. To see things, we must have ______________ .

5. A ______________ can bend light.

6. When light ______________ off objects and enters our eyes, we can see those objects.

7. If you shine light through a prism, you can see a ______________ .

© Macmillan/McGraw-Hill

LESSON
Cloze Activity

Name ______________________________ Date ____________

Light

Fill in the blanks. Use the words from the box.

colors	prism	shadow
light	reflects	solid

You would not be able to see anything if there were no light. Some sources of ______________ are the Sun, lightbulbs, and flashlights. We see objects because the light from these sources ______________ off of objects around us. A ______________ is a dark area that light does not reach. Light cannot pass through some ______________ objects. Light can pass through clear objects such as, glass.

Light is a mix of all ______________ . An object that makes light bend is called a ______________ . When light bends, it separates into the different colors of the rainbow.

© Macmillan/McGraw-Hill

Exploring Electricity

Use your book to help you fill in the blanks.

What is current electricity?

1. Electricity is a kind of ______________ that gives off light and heat.
2. Electricity that moves in a path is called ________________ .
3. We call this path a ______________ .
4. Current electricity can come from ______________ or from outlets.
5. Power ______________ make electricity that connects to wall outlets in homes.

What is static electricity?

6. The kind of energy that helps things stick together is called ________________ .

© Macmillan/McGraw-Hill

Name ____________________________ Date ____________

7. Pieces of matter push toward or pull from each other when they have a ______________ .

8. A charge can build up on one object and then ______________ to another object.

9. This is how ______________ works.

10. Charges build in storm ______________ and then jump to the ground.

Critical Thinking

11. How are a flashlight and lightning similar? How are they different?

____

____

____

____

____

____

____

© Macmillan/McGraw-Hill

Name ______________________________ Date ____________

It's Electric

Read the Reading in Science pages in your book. As you read, keep track of what happens and why. Record the causes and effects you read about in the chart below. Remember, a cause is why something happens. An effect is the thing that happens.

Cause	Effect
coal, oil, wind, water, or nuclear reactions	
	The generator creates electricity.
	Electricity flows from the power plant through power lines to your home.

© Macmillan/McGraw-Hill

Name ______________________ Date ____________

Energy is needed to make electricity. Where can that energy come from?

____

____

Where does energy come from in your community? Ask an adult to help you find out!

____

____

Write About It

Cause and Effect. How does electricity help make your life easier?

____

____

____

© Macmillan/McGraw-Hill